CONTINUUM MECHANICS

Concise Theory and Problems

Second Corrected and Enlarged Edition

P. CHADWICK

School of Mathematics and Physics
University of East Anglia, Norwich

DOVER PUBLICATIONS, INC.
Mineola, New York

Bibliographical Note

This Dover edition, first published in 1999, is a corrected and enlarged edition of the work originally published in 1976 by George Allen & Unwin Ltd., London. It contains four new appendices written by the author especially for the Dover edition.

Library of Congress Cataloging-in-Publication Data

Chadwick, Peter, 1931–
Continuum mechanics : concise theory and problems / P. Chadwick. — 2nd corr. and enl. ed.
p. cm.
Corrected and enlarged ed. of the work originally published in 1976 by George Allen & Unwin, Ltd., London.
Includes index.
ISBN-13: 978-0-486-40180-5 (pbk.)
ISBN-10: 0-486-40180-4 (pbk.)
1. Continuum mechanics. I. Title.

QA808.C46 1999
531—dc21

99-38303
CIP

Manufactured in the United States by Courier Corporation
40180405
www.doverpublications.com

CONTINUUM MECHANICS
Concise Theory and Problems

Preface

At the present time the number of universities and colleges offering courses on continuum mechanics is increasing and the advantages of presenting to students a unified basis for further work in fluid dynamics and the mechanics of solid materials are being more widely recognized by teachers of theoretical mechanics. A subject which is relatively new to degree syllabuses inevitably takes some time to receive adequate textbook coverage, and exercises suitable for use in class and in set work are apt initially to be in short supply. This book is an outgrowth of lecture courses which have been given over the past six years in the University of East Anglia to second- and final-year undergraduates specializing in mathematics and to first-year graduate students taking a master's degree course in theoretical mechanics. Colleagues engaged in the teaching of continuum mechanics have urged me to give wider currency to the sets of problems which have been compiled for use in tutorials and examples classes and in the examinations associated with these courses. This material appears here in an expanded form, and in view of the lack of suitable elementary textbooks to which readers can be referred for accounts of the underlying results, an explanatory text has been provided.

The book is intended primarily for use in conjunction with a lecture course, or equivalent form of teaching. The treatment of the principles of continuum mechanics, while reasonably complete mathematically, is, by design, concise; and to a considerable extent physical considerations, motivating arguments and detailed interpretations have been omitted. In matters of this kind the individual teacher will wish to exercise his own point of view: if the time at his disposal is increased by the availability to his students of this plain account of the theoretical groundwork, I shall be well satisfied. The emphasis throughout is on the mechanics of continuous media in general, and the book ends at a point from which courses specifically devoted to the mechanics of fluids and solid materials could logically proceed. Illustrative problems on linearly viscous fluids and isotropic elastic solids are given in the last two sections in order to provide links with traditional developments of these subjects.

The reader is assumed to have a knowledge of algebra and analysis which includes the theory of vector spaces and calculus in n-dimensional Euclidean space, and familiarity with particle and rigid body mechanics and the geometrical treatment of vectors is taken for granted. In Britain a student reading for honours in mathematics normally covers these topics in the first half of his course, while most students reading mathematical or theoretical physics, or taking a combined honours course, should be adequately prepared by the start of the final year. For applied science students the subject-matter is likely to be accessible only at post-graduate level.

In a book catering mainly for the needs of beginners in continuum mechanics I have not felt it appropriate to give references to the original sources of all the material presented. An advanced treatment of the matters discussed in these pages, with full historical notes and literature references, can be found in the authoritative works of Truesdell–Toupin and Truesdell–Noll†, and I wish to acknowledge here my indebtedness to the writings of these authors.

I shall be grateful for any suggestions or corrections from those who use this book.

P. C.

† C. Truesdell and R. A. Toupin, *The Classical Field Theories.* Handbuch der Physik (ed. S. Flügge) Vol. III/1, pp. 226–858 (Berlin etc., Springer, 1960).
C. Truesdell and W. Noll, *The Non-Linear Field Theories of Mechanics.* Handbuch der Physik (ed. S. Flügge) Vol. III/3 (Berlin etc., Springer, 1965).

Contents

Note to the reader

In each chapter equations to which subsequent reference is made are numbered from (1). Equation (43) of Chapter 1 is referred to as (43) *in that chapter*: in the later chapters it is referred to as (1.43). The same convention applies in references to sections, problems, exercises and figures.

Chapter 1

VECTOR AND TENSOR THEORY

The theory of scalar-, vector- and tensor-valued functions defined on subsets of a three-dimensional Euclidean space is a major part of the mathematical framework upon which continuum mechanics is built. This chapter is intended to provide a concise survey of basic results needed in the rest of the book and its contents will be found to be closely integrated into the subsequent text. It is not advisable, however, for the reader to postpone his study of continuum mechanics until the whole of this material has been mastered. Rather he should use Sections 1 to 3, 9 and 10 to refresh, and perhaps reorientate, his knowledge of vector algebra and analysis, and then turn back to the topics discussed in the remaining sections as the need arises.

1 VECTOR ALGEBRA

The scalar and vector products, with which the reader is already assumed to be familiar, can be defined by a system of axioms and we follow this approach here as a means of reviewing briefly the essential facts of vector algebra and, at the same time, providing a natural starting point for the development of tensor algebra. Contact with the geometrical viewpoint customarily adopted in elementary treatments of vector theory is made in Section 9.

Let E be a three-dimensional vector space over the field R of real numbers. We say that E is a *Euclidean vector space* if, to each pair of vectors $\boldsymbol{a}$, $\boldsymbol{b}$ in E, there corresponds a scalar (in R), denoted by $\boldsymbol{a}\,.\,\boldsymbol{b}$ and called the *scalar product* of $\boldsymbol{a}$ and $\boldsymbol{b}$, and a vector (in E), written $\boldsymbol{a} \wedge \boldsymbol{b}$ and referred to as the *vector product* of $\boldsymbol{a}$ and $\boldsymbol{b}$, with the following properties:

$$\boldsymbol{b}\,.\,\boldsymbol{a} = \boldsymbol{a}\,.\,\boldsymbol{b} \quad \forall\, \boldsymbol{a}, \boldsymbol{b} \in \mathsf{E}, \tag{1}$$

$$(\alpha\boldsymbol{a} + \beta\boldsymbol{b})\,.\,\boldsymbol{c} = \alpha(\boldsymbol{a}\,.\,\boldsymbol{c}) + \beta(\boldsymbol{b}\,.\,\boldsymbol{c}) \quad \forall\, \boldsymbol{a}, \boldsymbol{b}, \boldsymbol{c} \in \mathsf{E}, \alpha, \beta \in \mathsf{R}, \tag{2}$$

$$\boldsymbol{a}\,.\,\boldsymbol{a} \geqslant 0 \quad \forall\, \boldsymbol{a} \in \mathsf{E} \text{ with } \boldsymbol{a}\,.\,\boldsymbol{a} = 0 \text{ if and only if } \boldsymbol{a} = \boldsymbol{0}, \tag{3}$$

$$\boldsymbol{b} \wedge \boldsymbol{a} = -\boldsymbol{a} \wedge \boldsymbol{b} \quad \forall\, \boldsymbol{a}, \boldsymbol{b} \in \mathsf{E}, \tag{4}$$

$$(\alpha\boldsymbol{a} + \beta\boldsymbol{b}) \wedge \boldsymbol{c} = \alpha(\boldsymbol{a} \wedge \boldsymbol{c}) + \beta(\boldsymbol{b} \wedge \boldsymbol{c}) \quad \forall\, \boldsymbol{a}, \boldsymbol{b}, \boldsymbol{c} \in \mathsf{E}, \alpha, \beta \in \mathsf{R}, \tag{5}$$

$$\boldsymbol{a}\,.\,(\boldsymbol{a} \wedge \boldsymbol{b}) = 0 \quad \forall\, \boldsymbol{a}, \boldsymbol{b} \in \mathsf{E}, \tag{6}$$

$$(\boldsymbol{a} \wedge \boldsymbol{b})\,.\,(\boldsymbol{a} \wedge \boldsymbol{b}) = (\boldsymbol{a}\,.\,\boldsymbol{a})(\boldsymbol{b}\,.\,\boldsymbol{b}) - (\boldsymbol{a}\,.\,\boldsymbol{b})^2 \quad \forall\, \boldsymbol{a}, \boldsymbol{b} \in \mathsf{E}. \tag{7}$$

The *norm* (or *magnitude*), $|\boldsymbol{a}|$, of a vector $\boldsymbol{a}$ is defined by

$$|\boldsymbol{a}| = (\boldsymbol{a}\,.\,\boldsymbol{a})^{1/2}, \tag{8}$$

and a vector with unit norm is termed a *unit vector*. Two vectors $\boldsymbol{a}$ and $\boldsymbol{b}$ are said to be *orthogonal* if $\boldsymbol{a}\,.\,\boldsymbol{b} = 0$.

Problem 1 Prove that $\boldsymbol{a} \wedge \boldsymbol{b} = \boldsymbol{0}$ if and only if $\boldsymbol{a}$ and $\boldsymbol{b}$ are linearly dependent.

Solution. (*a*) If $\boldsymbol{a}$ and $\boldsymbol{b}$ are linearly dependent, either $\boldsymbol{a} = \boldsymbol{0}$ or there is a scalar α such that $\boldsymbol{b} = \alpha\boldsymbol{a}$. In the first case it follows from axiom (5) that $\boldsymbol{a} \wedge \boldsymbol{b} = \boldsymbol{0}$, and in the second case the same conclusion is reached via axioms (4) and (5).

(*b*) If $\boldsymbol{a} \wedge \boldsymbol{b} = \boldsymbol{0}$, axioms (7) and (3) and equation (8) give $\boldsymbol{a}\,.\,\boldsymbol{b} = \pm|\boldsymbol{a}|\,|\boldsymbol{b}|$. Suppose that the plus sign applies. Then, using axioms (2) and (1),

$$(|\boldsymbol{b}|\boldsymbol{a} - |\boldsymbol{a}|\boldsymbol{b})\,.\,(|\boldsymbol{b}|\boldsymbol{a} - |\boldsymbol{a}|\boldsymbol{b}) = 2|\boldsymbol{a}|^2|\boldsymbol{b}|^2 - 2|\boldsymbol{a}|\,|\boldsymbol{b}|\boldsymbol{a}\,.\,\boldsymbol{b} = 0,$$

and, in consequence of (3), $|\boldsymbol{b}|\boldsymbol{a} = |\boldsymbol{a}|\boldsymbol{b}$. When the minus sign holds, $|\boldsymbol{b}|\boldsymbol{a} = -|\boldsymbol{a}|\boldsymbol{b}$ by a similar argument. In both cases either $\boldsymbol{a} = \boldsymbol{0}$ or $\boldsymbol{b}$ is a scalar multiple of $\boldsymbol{a}$. Thus $\boldsymbol{a}$ and $\boldsymbol{b}$ are linearly dependent.

The *scalar triple product* of three vectors $\boldsymbol{a}$, $\boldsymbol{b}$, $\boldsymbol{c}$, denoted by $[\boldsymbol{a}, \boldsymbol{b}, \boldsymbol{c}]$, is defined by

$$[\boldsymbol{a}, \boldsymbol{b}, \boldsymbol{c}] = \boldsymbol{a}\,.\,(\boldsymbol{b} \wedge \boldsymbol{c}). \tag{9}$$

Henceforth the abbreviated term *triple product* is used.

Problem 2 Establish the following properties of the triple product:

$$\text{(i) } [\boldsymbol{a}, \boldsymbol{b}, \boldsymbol{c}] = [\boldsymbol{b}, \boldsymbol{c}, \boldsymbol{a}] = [\boldsymbol{c}, \boldsymbol{a}, \boldsymbol{b}] = -[\boldsymbol{a}, \boldsymbol{c}, \boldsymbol{b}] = -[\boldsymbol{b}, \boldsymbol{a}, \boldsymbol{c}]$$
$$= -[\boldsymbol{c}, \boldsymbol{b}, \boldsymbol{a}] \quad \forall\, \boldsymbol{a}, \boldsymbol{b}, \boldsymbol{c} \in \mathsf{E}, \tag{10}$$

(ii) $[\alpha \boldsymbol{a} + \beta \boldsymbol{b}, \boldsymbol{c}, \boldsymbol{d}] = \alpha[\boldsymbol{a}, \boldsymbol{c}, \boldsymbol{d}] + \beta[\boldsymbol{b}, \boldsymbol{c}, \boldsymbol{d}]$

$$\forall\, \boldsymbol{a}, \boldsymbol{b}, \boldsymbol{c}, \boldsymbol{d} \in \mathsf{E},\ \alpha, \beta \in \mathsf{R}, \qquad (11)$$

(iii) $[\boldsymbol{a}, \boldsymbol{b}, \boldsymbol{c}] = 0$ if and only if $\boldsymbol{a}, \boldsymbol{b}, \boldsymbol{c}$ are linearly dependent.

Solution. (i) In view of axioms (4), (2) and (1) the sign of a triple product is reversed when the second and third members of the product are exchanged. From axiom (6), with appeal to (2) and (5),

$$\begin{aligned} 0 &= (\boldsymbol{a} + \boldsymbol{b}) \,.\, \{(\boldsymbol{a} + \boldsymbol{b}) \wedge \boldsymbol{c}\} \\ &= \boldsymbol{a} \,.\, (\boldsymbol{a} \wedge \boldsymbol{c}) + \boldsymbol{a} \,.\, (\boldsymbol{b} \wedge \boldsymbol{c}) + \boldsymbol{b} \,.\, (\boldsymbol{a} \wedge \boldsymbol{c}) + \boldsymbol{b} \,.\, (\boldsymbol{b} \wedge \boldsymbol{c}) \\ &= [\boldsymbol{a}, \boldsymbol{b}, \boldsymbol{c}] + [\boldsymbol{b}, \boldsymbol{a}, \boldsymbol{c}], \end{aligned}$$

whence it is clear that sign reversal also results from interchanging the first and second members of a triple product. Repeated application of these properties leads to the identities (10).

(ii) Replace $\boldsymbol{c}$ by $\boldsymbol{c} \wedge \boldsymbol{d}$ in axiom (2).

(iii) (*a*) First observe that axiom (2), in conjunction with the properties (10), implies that a triple product having the zero vector as one of its members vanishes. If $\boldsymbol{a}, \boldsymbol{b}, \boldsymbol{c}$ are linearly dependent there exist scalars α, β, γ, not all zero, such that $\alpha\boldsymbol{a} + \beta\boldsymbol{b} + \gamma\boldsymbol{c} = \boldsymbol{0}$. Hence the triple products

$$[\alpha\boldsymbol{a} + \beta\boldsymbol{b} + \gamma\boldsymbol{c}, \boldsymbol{b}, \boldsymbol{c}], \quad [\boldsymbol{a}, \alpha\boldsymbol{a} + \beta\boldsymbol{b} + \gamma\boldsymbol{c}, \boldsymbol{c}], \quad [\boldsymbol{a}, \boldsymbol{b}, \alpha\boldsymbol{a} + \beta\boldsymbol{b} + \gamma\boldsymbol{c}]$$

are all zero and, with the use of equations (10) and (11), they reduce in turn to $\alpha[\boldsymbol{a}, \boldsymbol{b}, \boldsymbol{c}]$, $\beta[\boldsymbol{a}, \boldsymbol{b}, \boldsymbol{c}]$, $\gamma[\boldsymbol{a}, \boldsymbol{b}, \boldsymbol{c}]$. Since at least one of α, β, γ is non-zero, $[\boldsymbol{a}, \boldsymbol{b}, \boldsymbol{c}] = 0$.

(*b*) The converse result is proved by contradiction. Suppose that $[\boldsymbol{a}, \boldsymbol{b}, \boldsymbol{c}] = 0$ and that $\boldsymbol{a}, \boldsymbol{b}, \boldsymbol{c}$ are linearly independent vectors. Problem 1 tells us that $\boldsymbol{b} \wedge \boldsymbol{c} \neq \boldsymbol{0}$, and we see from equations (9) and (10) that $\boldsymbol{a}$, $\boldsymbol{b}$ and $\boldsymbol{c}$ are each orthogonal to $\boldsymbol{b} \wedge \boldsymbol{c}$. Since $\boldsymbol{a}, \boldsymbol{b}, \boldsymbol{c}$ form a basis of E it follows that every vector is orthogonal to $\boldsymbol{b} \wedge \boldsymbol{c}$. This conclusion is plainly false, however, as $\boldsymbol{b} \wedge \boldsymbol{c}$ is not orthogonal to itself, so $\boldsymbol{a}, \boldsymbol{b}, \boldsymbol{c}$ must be linearly dependent.

A fundamental property of a finite-dimensional vector space equipped with a scalar product is the existence of an orthonormal basis.[1] Thus, in the Euclidean vector space **E** defined above there is a set of three vectors, $e = \{\boldsymbol{e}_1, \boldsymbol{e}_2, \boldsymbol{e}_3\}$, such that

$$\boldsymbol{e}_i \cdot \boldsymbol{e}_j = \delta_{ij} \quad (i, j = 1, 2, 3), \tag{12}$$

where δ_{ij}, the *Kronecker delta*, takes the value 1 when $i = j$ and 0 when $i \neq j$. Corresponding to an arbitrary vector $\boldsymbol{a}$ there is an ordered triplet of scalars, (a_1, a_2, a_3), such that

$$\boldsymbol{a} = a_1\boldsymbol{e}_1 + a_2\boldsymbol{e}_2 + a_3\boldsymbol{e}_3 = a_p\boldsymbol{e}_p; \tag{13}$$

a_1, a_2, a_3 are called the *components of* $\boldsymbol{a}$ relative to the basis e. On the right of (13) use has been made of the *summation convention* according to which any expression in which a suffix appears twice is understood to be summed over the range, 1, 2, 3 of that suffix. *Only the letters p, q, r, s and π, ρ, σ are used in this book for repeated subscripts and summation over indices appearing more than twice is indicated explicitly.* In view of the orthonormality condition (12),

$$\boldsymbol{a} \cdot \boldsymbol{e}_i = a_p\boldsymbol{e}_p \cdot \boldsymbol{e}_i = a_p\delta_{pi} = a_i,$$

whence

$$\boldsymbol{a} = (\boldsymbol{a} \cdot \boldsymbol{e}_p)\boldsymbol{e}_p. \tag{14}$$

Problem 3 Show that

$$\boldsymbol{e}_2 \wedge \boldsymbol{e}_3 = \pm\boldsymbol{e}_1, \quad \boldsymbol{e}_3 \wedge \boldsymbol{e}_1 = \pm\boldsymbol{e}_2, \quad \boldsymbol{e}_1 \wedge \boldsymbol{e}_2 = \pm\boldsymbol{e}_3, \tag{15}$$

the plus or minus signs holding together.

Solution. On setting $\boldsymbol{a} = \boldsymbol{e}_2 \wedge \boldsymbol{e}_3$ in equation (14) we obtain

$$\boldsymbol{e}_2 \wedge \boldsymbol{e}_3 = \{(\boldsymbol{e}_2 \wedge \boldsymbol{e}_3) \cdot \boldsymbol{e}_p\}\boldsymbol{e}_p = [\boldsymbol{e}_2, \boldsymbol{e}_3, \boldsymbol{e}_p]\boldsymbol{e}_p = [\boldsymbol{e}_1, \boldsymbol{e}_2, \boldsymbol{e}_3]\boldsymbol{e}_1, \tag{A}$$

use being made of equations (9) and (10). Similarly,

$$\boldsymbol{e}_3 \wedge \boldsymbol{e}_1 = [\boldsymbol{e}_1, \boldsymbol{e}_2, \boldsymbol{e}_3]\boldsymbol{e}_2, \quad \boldsymbol{e}_1 \wedge \boldsymbol{e}_2 = [\boldsymbol{e}_1, \boldsymbol{e}_2, \boldsymbol{e}_3]\boldsymbol{e}_3. \tag{B}$$

With the aid of (12) we deduce from (A) and (B) that

$$|\boldsymbol{e}_2 \wedge \boldsymbol{e}_3|^2 = |\boldsymbol{e}_3 \wedge \boldsymbol{e}_1|^2 = |\boldsymbol{e}_1 \wedge \boldsymbol{e}_2|^2 = [\boldsymbol{e}_1, \boldsymbol{e}_2, \boldsymbol{e}_3]^2,$$

while axiom (7), in combination with (12), yields the same results with

[1] See, for example, K. Hoffman and R. Kunze, *Linear Algebra* (Englewood Cliffs, N. J., Prentice-Hall, 1961), p. 230.

$[\boldsymbol{e}_1, \boldsymbol{e}_2, \boldsymbol{e}_3]^2$ replaced by 1. Hence $[\boldsymbol{e}_1, \boldsymbol{e}_2, \boldsymbol{e}_3] = \pm 1$ and equations (15) follow directly from (A) and (B).

The relations (15) are contained in the compact expression

$$\boldsymbol{e}_i \wedge \boldsymbol{e}_j = \pm \varepsilon_{ijp} \boldsymbol{e}_p, \tag{16}$$

where ε_{ijk}, the *alternator*, takes the value 1 when i, j, k is a cyclic permutation of 1, 2, 3, -1 when i, j, k is a non-cyclic permutation of 1, 2, 3, and is otherwise zero.

Let $\boldsymbol{a}, \boldsymbol{b}$ be an arbitrary pair of vectors having components a_i, b_i relative to e. Then, with the use of axioms (1) and (2) and equation (12),

$$\boldsymbol{a} \cdot \boldsymbol{b} = (a_p \boldsymbol{e}_p) \cdot (b_q \boldsymbol{e}_q) = a_p b_q \boldsymbol{e}_p \cdot \boldsymbol{e}_q = a_p b_q \delta_{pq} = a_p b_p. \tag{17}$$

This formula for the scalar product of two vectors in terms of their components relative to an orthonormal basis confirms that there is precisely one scalar product on E satisfying axioms (1) to (3). When $\boldsymbol{b} = \boldsymbol{a}$, equations (17) and (8) give

$$|\boldsymbol{a}| = (a_p a_p)^{1/2}. \tag{18}$$

The component form of the vector product $\boldsymbol{a} \wedge \boldsymbol{b}$ relative to the basis e is found with the aid of axioms (4) and (5) and equation (16) to be

$$\boldsymbol{a} \wedge \boldsymbol{b} = (a_p \boldsymbol{e}_p) \wedge (b_q \boldsymbol{e}_q) = a_p b_q \boldsymbol{e}_p \wedge \boldsymbol{e}_q = \pm \varepsilon_{pqr} a_p b_q \boldsymbol{e}_r. \tag{19}$$

It is seen from (19) that there are two vector products on E obeying the axioms (4) to (7), one being the negative of the other. If $\boldsymbol{c}$ is a third arbitrary vector with components c_i relative to e, the component form of the triple product $[\boldsymbol{a}, \boldsymbol{b}, \boldsymbol{c}]$, obtained by combining equation (9) with (17) and (19), is

$$[\boldsymbol{a}, \boldsymbol{b}, \boldsymbol{c}] = \pm \varepsilon_{pqr} a_p b_q c_r. \tag{20}$$

Two ordered bases of E are said to be *similar* if their triple products have the same sign.[2] Similarity is an equivalence relation which partitions the collection of all ordered bases of E into two classes, one containing the members with positive triple products and the other those with negative triple products. The pairing of E with a

[2] Property (iii) of Problem 2 (p. 12) implies that the triple product of three vectors forming a basis of E is non-zero.

rule[3] identifying the equivalence class to which each of its ordered bases belongs is called an *orientation* of E. There are two such orientations, denoted by E^+ and E^-, and an ordered basis of E is said to be *positive in* E^+ if its triple product is positive and *positive in* E^- if its triple product is negative. The conferment of an orientation on E removes the ambiguity of sign in equations (15), (16), (19) and (20), the upper or lower sign being chosen according as e is positive in E^+ or E^-. Evidently an *orientated* Euclidean vector space has a unique vector product meeting the axioms (4) to (7).

2 TENSOR ALGEBRA

A *tensor*[4] $\boldsymbol{A}$ is a linear transformation of the Euclidean vector space E into itself. Specifically, $\boldsymbol{A}$ assigns to an arbitrary vector $\boldsymbol{a}$ a vector, denoted by $\boldsymbol{Aa}$, in such a way that

$$\boldsymbol{A}(\alpha\boldsymbol{a} + \beta\boldsymbol{b}) = \alpha(\boldsymbol{Aa}) + \beta(\boldsymbol{Ab}) \quad \forall\, \boldsymbol{a}, \boldsymbol{b} \in \mathsf{E}, \alpha, \beta \in \mathsf{R}. \tag{21}$$

The set of all tensors on E is denoted by L.

Two tensors are equal if and only if their actions on an arbitrary vector are identical, and the rules for the addition, scalar multiplication and multiplication (or composition) of tensors are

$$\left.\begin{aligned} (\boldsymbol{A} + \boldsymbol{B})\boldsymbol{a} &= \boldsymbol{Aa} + \boldsymbol{Ba} \quad \forall\, \boldsymbol{A}, \boldsymbol{B} \in \mathsf{L}, \boldsymbol{a} \in \mathsf{E},\\ (\alpha\boldsymbol{A})\boldsymbol{a} &= \alpha(\boldsymbol{Aa}) \quad \forall\, \boldsymbol{A} \in \mathsf{L}, \boldsymbol{a} \in \mathsf{E}, \alpha \in \mathsf{R},\\ (\boldsymbol{AB})\boldsymbol{a} &= \boldsymbol{A}(\boldsymbol{Ba}) \quad \forall\, \boldsymbol{A}, \boldsymbol{B} \in \mathsf{L}, \boldsymbol{a} \in \mathsf{E}. \end{aligned}\right\} \tag{22}$$

The *zero tensor* $\boldsymbol{O}$ assigns to $\boldsymbol{a}$ the zero vector and the *identity tensor* $\boldsymbol{I}$ assigns to $\boldsymbol{a}$ the vector $\boldsymbol{a}$ itself:

$$\boldsymbol{Oa} = \boldsymbol{0}, \quad \boldsymbol{Ia} = \boldsymbol{a} \quad \forall\, \boldsymbol{a} \in \mathsf{E}. \tag{23}$$

We leave it to the reader to establish, on the basis of equations (21) to (23), the following properties:

(i) $\boldsymbol{A} + \boldsymbol{B} = \boldsymbol{B} + \boldsymbol{A} \quad \forall\, \boldsymbol{A}, \boldsymbol{B} \in \mathsf{L}$,

(ii) $\alpha(\boldsymbol{AB}) = (\alpha\boldsymbol{A})\boldsymbol{B} = \boldsymbol{A}(\alpha\boldsymbol{B}) \quad \forall\, \boldsymbol{A}, \boldsymbol{B} \in \mathsf{L}, \alpha \in \mathsf{R}$,

[3] For example, the statement that $[\boldsymbol{e}_1, \boldsymbol{e}_2, \boldsymbol{e}_3] = 1$ for a particular orthonormal basis.

[4] Strictly, $\boldsymbol{A}$ is a *second-order* tensor, but we shall have no occasion in this book to introduce tensors of order higher than two.

$$\left.\begin{aligned} &\text{(iii) } \boldsymbol{A}(\boldsymbol{B}+\boldsymbol{C}) = \boldsymbol{AB}+\boldsymbol{AC} \\ &\qquad (\boldsymbol{A}+\boldsymbol{B})\boldsymbol{C} = \boldsymbol{AC}+\boldsymbol{BC} \\ &\text{(iv) } \boldsymbol{A}(\boldsymbol{BC}) = (\boldsymbol{AB})\boldsymbol{C} \end{aligned}\right\} \ \forall\, \boldsymbol{A},\boldsymbol{B},\boldsymbol{C} \in \mathsf{L},$$

$$\text{(v) } \boldsymbol{A}\mathbf{O} = \mathbf{O}\boldsymbol{A} = \mathbf{O}, \quad \boldsymbol{AI} = \boldsymbol{IA} = \boldsymbol{A} \quad \forall\, \boldsymbol{A} \in \mathsf{L}.$$

Associated with an arbitrary tensor $\boldsymbol{A}$ there is a unique tensor $\boldsymbol{A}^{\mathrm{T}}$, called the *transpose* of $\boldsymbol{A}$, such that[5]

$$\boldsymbol{a}\,.\,(\boldsymbol{A}^{\mathrm{T}}\boldsymbol{b}) = \boldsymbol{b}\,.\,(\boldsymbol{A}\boldsymbol{a}) \quad \forall\, \boldsymbol{a}, \boldsymbol{b} \in \mathsf{E}. \tag{24}$$

It follows from this definition that $(\boldsymbol{A}^{\mathrm{T}})^{\mathrm{T}} = \boldsymbol{A}$ and also, with the use of equations (22) and axioms (1) and (2), that

$$(\alpha\boldsymbol{A} + \beta\boldsymbol{B})^{\mathrm{T}} = \alpha\boldsymbol{A}^{\mathrm{T}} + \beta\boldsymbol{B}^{\mathrm{T}} \quad \forall\, \boldsymbol{A}, \boldsymbol{B} \in \mathsf{L}, \alpha, \beta \in \mathsf{R}, \tag{25}$$

$$(\boldsymbol{AB})^{\mathrm{T}} = \boldsymbol{B}^{\mathrm{T}}\boldsymbol{A}^{\mathrm{T}} \quad \forall\, \boldsymbol{A}, \boldsymbol{B} \in \mathsf{L}. \tag{26}$$

A tensor $\boldsymbol{A}$ such that $\boldsymbol{A}^{\mathrm{T}} = \boldsymbol{A}$ is said to be *symmetric*, and if $\boldsymbol{A}^{\mathrm{T}} = -\boldsymbol{A}$, $\boldsymbol{A}$ is called a *skew-symmetric* tensor. The identity

$$\boldsymbol{A} = \tfrac{1}{2}(\boldsymbol{A} + \boldsymbol{A}^{\mathrm{T}}) + \tfrac{1}{2}(\boldsymbol{A} - \boldsymbol{A}^{\mathrm{T}}) \tag{27}$$

demonstrates that an arbitrary tensor can be expressed as the sum of symmetric and skew-symmetric parts. This decomposition is unique.

Problem 4 Let $\{\boldsymbol{f}, \boldsymbol{g}, \boldsymbol{h}\}$ and $\{\boldsymbol{l}, \boldsymbol{m}, \boldsymbol{n}\}$ be arbitrarily chosen bases of E and let $\boldsymbol{A}$ be an arbitrary tensor. Show that

$$\begin{aligned} &\{[\boldsymbol{Af}, \boldsymbol{g}, \boldsymbol{h}] + [\boldsymbol{f}, \boldsymbol{Ag}, \boldsymbol{h}] + [\boldsymbol{f}, \boldsymbol{g}, \boldsymbol{Ah}]\}/[\boldsymbol{f}, \boldsymbol{g}, \boldsymbol{h}] \\ &\qquad = \{[\boldsymbol{Al}, \boldsymbol{m}, \boldsymbol{n}] + [\boldsymbol{l}, \boldsymbol{Am}, \boldsymbol{n}] + [\boldsymbol{l}, \boldsymbol{m}, \boldsymbol{An}]\}/[\boldsymbol{l}, \boldsymbol{m}, \boldsymbol{n}], \end{aligned} \tag{A}$$

$$\begin{aligned} &\{[\boldsymbol{f}, \boldsymbol{Ag}, \boldsymbol{Ah}] + [\boldsymbol{Af}, \boldsymbol{g}, \boldsymbol{Ah}] + [\boldsymbol{Af}, \boldsymbol{Ag}, \boldsymbol{h}]\}/[\boldsymbol{f}, \boldsymbol{g}, \boldsymbol{h}] \\ &\qquad = \{[\boldsymbol{l}, \boldsymbol{Am}, \boldsymbol{An}] + [\boldsymbol{Al}, \boldsymbol{m}, \boldsymbol{An}] + [\boldsymbol{Al}, \boldsymbol{Am}, \boldsymbol{n}]\}/[\boldsymbol{l}, \boldsymbol{m}, \boldsymbol{n}], \end{aligned} \tag{B}$$

$$[\boldsymbol{Af}, \boldsymbol{Ag}, \boldsymbol{Ah}]/[\boldsymbol{f}, \boldsymbol{g}, \boldsymbol{h}] = [\boldsymbol{Al}, \boldsymbol{Am}, \boldsymbol{An}]/[\boldsymbol{l}, \boldsymbol{m}, \boldsymbol{n}]. \tag{C}$$

Solution. Let f_i, g_i, h_i be the components of $\boldsymbol{f}$, $\boldsymbol{g}$, $\boldsymbol{h}$ relative to the orthonormal basis *e*. Then, using equations (13), (21), (10) and (11),

[5] See K. Hoffman and R. Kunze, *op. cit.* p. 237.

we find that

$$[Af, g, h] = [A(f_p e_p), g_q e_q, h_r e_r]$$
$$= [f_p A e_p, g_q e_q, h_r e_r] = f_p g_q h_r [A e_p, e_q, e_r],$$

and it follows that the left-hand side of (A) can be put into the form

$$f_p g_q h_r \{[Ae_p, e_q, e_r] + [e_p, Ae_q, e_r] + [e_p, e_q, Ae_r]\}/[f, g, h]. \quad \text{(D)}$$

The expression in curly brackets is unchanged by a cyclic permutation of the suffixes p, q, r; its sign is reversed by a non-cyclic permutation of these subscripts; and if any two of p, q, r have equal values the expression is zero. Recalling the definition of the alternator and equation (20), we can therefore rewrite (D) as

$$f_p g_q h_r \varepsilon_{pqr} \{[Ae_1, e_2, e_3] + [e_1, Ae_2, e_3] + [e_1, e_2, Ae_3]\}/[f, g, h]$$
$$= \pm\{[Ae_1, e_2, e_3] + [e_1, Ae_2, e_3] + [e_1, e_2, Ae_3]\}. \quad \text{(E)}$$

If e is positive in E^+, the upper sign applies in (E) and $[e_1, e_2, e_3] = 1$; if e is positive in E^-, the lower sign must be chosen and $[e_1, e_2, e_3] = -1$. Thus, regardless of orientation,

$$\{[Af, g, h] + [f, Ag, h] + [f, g, Ah]\}/[f, g, h]$$
$$= \{[Ae_1, e_2, e_3] + [e_1, Ae_2, e_3] + [e_1, e_2, Ae_3]\}/[e_1, e_2, e_3], \quad \text{(F)}$$

and since the choice of basis on the left of (F) is arbitrary, we arrive immediately at the result (A).

The steps which have been used to establish (A) also yield derivations of equations (B) and (C) and the reader should check his understanding of the solution by writing out the details.

It is a consequence of Problem 4 that, corresponding to an arbitrary tensor A, there are scalars, I_A, II_A, III_A such that

$$\left.\begin{aligned} &[Aa, b, c] + [a, Ab, c] + [a, b, Ac] = I_A[a, b, c] && (28)\\ &[a, Ab, Ac] + [Aa, b, Ac] + [Aa, Ab, c] = II_A[a, b, c] && (29)\\ &[Aa, Ab, Ac] = III_A[a, b, c] && (30)\end{aligned}\right\} \forall a, b, c \in \mathsf{E}.$$

These scalars are called the *principal invariants of* A, but I_A is more commonly referred to as the *trace of* $\boldsymbol{A}$, written tr $\boldsymbol{A}$, and III_A as the *determinant* of $\boldsymbol{A}$, denoted by det $\boldsymbol{A}$. Thus

$$I_A = \operatorname{tr} \boldsymbol{A}, \quad III_A = \det \boldsymbol{A}.$$

From equations $(22)_{1,2}$, with (11) and (10), we see that

$$\operatorname{tr}(\alpha \boldsymbol{A} + \beta \boldsymbol{B}) = \alpha \operatorname{tr} \boldsymbol{A} + \beta \operatorname{tr} \boldsymbol{B} \quad \forall\, \boldsymbol{A}, \boldsymbol{B} \in \mathsf{L}, \alpha, \beta \in \mathsf{R}, \tag{31}$$

the trace therefore being a linear function from L to R. Immediate consequences of the definition (30) are

$$\det(\alpha \boldsymbol{A}) = \alpha^3 \det \boldsymbol{A} \qquad \forall\, \boldsymbol{A} \in \mathsf{L}, \alpha \in \mathsf{R}, \tag{32}$$

$$\det(\boldsymbol{AB}) = \det \boldsymbol{A} \det \boldsymbol{B} \quad \forall\, \boldsymbol{A}, \boldsymbol{B} \in \mathsf{L}, \tag{33}$$

$$\det \boldsymbol{I} = 1. \tag{34}$$

It also follows from (30) that a tensor preserves the orientation of E if and only if its determinant is positive.

Problem 5 Given a tensor $\boldsymbol{A}$, show that there exists a non-zero vector $\boldsymbol{n}$ such that $\boldsymbol{An} = \boldsymbol{0}$ if and only if det $\boldsymbol{A} = 0$.

Solution. (*a*) Suppose that det $\boldsymbol{A} = III_A = 0$ and let $\{\boldsymbol{f}, \boldsymbol{g}, \boldsymbol{h}\}$ be an arbitrary basis of E. Then, from (30) and property (iii) of Problem 2 (p. 12), the vectors $\boldsymbol{Af}, \boldsymbol{Ag}, \boldsymbol{Ah}$ are linearly dependent. This means that there are scalars α, β, γ, not all zero, such that

$$\alpha(\boldsymbol{Af}) + \beta(\boldsymbol{Ag}) + \gamma(\boldsymbol{Ah}) = \boldsymbol{A}(\alpha \boldsymbol{f} + \beta \boldsymbol{g} + \gamma \boldsymbol{h}) = \boldsymbol{0},$$

use being made of (21). Thus $\boldsymbol{An} = \boldsymbol{0}$ where $\boldsymbol{n} = \alpha \boldsymbol{f} + \beta \boldsymbol{g} + \gamma \boldsymbol{h} \neq \boldsymbol{0}$.

(*b*) Conversely, if there is a non-zero $\boldsymbol{n}$ such that $\boldsymbol{An} = \boldsymbol{0}$, on choosing vectors $\boldsymbol{l}$, $\boldsymbol{m}$ which form with $\boldsymbol{n}$ a linearly independent set and then replacing $\{\boldsymbol{a}, \boldsymbol{b}, \boldsymbol{c}\}$ by $\{\boldsymbol{l}, \boldsymbol{m}, \boldsymbol{n}\}$ in equation (30), we conclude, with further appeal to property (iii) of Problem 2, that $III_A = \det \boldsymbol{A} = 0$.

If det $\boldsymbol{A} \neq 0$, $\boldsymbol{A}$ is said to be *invertible* since there then exists[6] a unique tensor, called the *inverse* of $\boldsymbol{A}$ and denoted by $\boldsymbol{A}^{-1}$, such that

$$\boldsymbol{AA}^{-1} = \boldsymbol{A}^{-1}\boldsymbol{A} = \boldsymbol{I}. \tag{35}$$

Equations (33) to (35) together give

$$\det \boldsymbol{A}^{-1} = (\det \boldsymbol{A})^{-1}, \tag{36}$$

[6] See K. Hoffman and R. Kunze, *op. cit.* p. 150.

while if $\boldsymbol{B}$ is a second invertible tensor it follows from (35) that $\boldsymbol{AB}$ is invertible and

$$(\boldsymbol{AB})^{-1} = \boldsymbol{B}^{-1}\boldsymbol{A}^{-1}. \tag{37}$$

Corresponding to an arbitrary tensor $\boldsymbol{A}$ there is a unique tensor $\boldsymbol{A}^*$, called the *adjugate* of $\boldsymbol{A}$, such that[7]

$$\boldsymbol{A}^*(\boldsymbol{a} \wedge \boldsymbol{b}) = (\boldsymbol{A}\boldsymbol{a}) \wedge (\boldsymbol{A}\boldsymbol{b}) \quad \forall \boldsymbol{a}, \boldsymbol{b} \in \mathsf{E}. \tag{38}$$

Suppose that $\boldsymbol{A}$ is invertible and that $\boldsymbol{a}, \boldsymbol{b}, \boldsymbol{c}$ are arbitrary vectors. Then

$$\begin{aligned}\{\boldsymbol{A}^{\mathrm{T}*}(\boldsymbol{a} \wedge \boldsymbol{b})\} \,.\, \boldsymbol{c} &= [\boldsymbol{A}^{\mathrm{T}}\boldsymbol{a}, \boldsymbol{A}^{\mathrm{T}}\boldsymbol{b}, \boldsymbol{A}^{\mathrm{T}}(\boldsymbol{A}^{-1})^{\mathrm{T}}\boldsymbol{c}] \\ &= (\det \boldsymbol{A}^{\mathrm{T}})\,(\boldsymbol{a} \wedge \boldsymbol{b}) \,.\, \{(\boldsymbol{A}^{-1})^{\mathrm{T}}\boldsymbol{c}\} \\ &= (\det \boldsymbol{A}^{\mathrm{T}})\,\{\boldsymbol{A}^{-1}(\boldsymbol{a} \wedge \boldsymbol{b})\} \,.\, \boldsymbol{c},\end{aligned}$$

where successive use has been made of equations (38), (35), (26), (30) and (24). Because of the arbitrariness of $\boldsymbol{a}$, $\boldsymbol{b}$, $\boldsymbol{c}$ there follows the connection

$$\boldsymbol{A}^{-1} = (\det \boldsymbol{A}^{\mathrm{T}})^{-1}\boldsymbol{A}^{\mathrm{T}*} \tag{39}$$

between the inverse of $\boldsymbol{A}$ and the adjugate of $\boldsymbol{A}^{\mathrm{T}}$.

Problem 6 Let $\boldsymbol{A}$ be an invertible tensor which depends upon a real parameter τ. Assuming that $\mathrm{d}\boldsymbol{A}/\mathrm{d}\tau$ exists, prove that

$$\frac{\mathrm{d}}{\mathrm{d}\tau}(\det \boldsymbol{A}) = (\det \boldsymbol{A}) \operatorname{tr}\left(\frac{\mathrm{d}\boldsymbol{A}}{\mathrm{d}\tau}\boldsymbol{A}^{-1}\right). \tag{40}$$

Solution. The result of differentiating each side of equation (30) with respect to τ is

$$\begin{aligned}&[\boldsymbol{a}, \boldsymbol{b}, \boldsymbol{c}]\frac{\mathrm{d}}{\mathrm{d}\tau}(\det \boldsymbol{A}) \\ &\qquad = [\boldsymbol{BAa}, \boldsymbol{Ab}, \boldsymbol{Ac}] + [\boldsymbol{Aa}, \boldsymbol{BAb}, \boldsymbol{Ac}] + [\boldsymbol{Aa}, \boldsymbol{Ab}, \boldsymbol{BAc}],\end{aligned}$$

where $\boldsymbol{B} = (\mathrm{d}\boldsymbol{A}/\mathrm{d}\tau)\boldsymbol{A}^{-1}$. In view of equations (28) and (30) this becomes

$$[\boldsymbol{a}, \boldsymbol{b}, \boldsymbol{c}]\frac{\mathrm{d}}{\mathrm{d}\tau}(\det \boldsymbol{A}) = \operatorname{tr} \boldsymbol{B}[\boldsymbol{Aa}, \boldsymbol{Ab}, \boldsymbol{Ac}] = (\operatorname{tr} \boldsymbol{B})(\det \boldsymbol{A})[\boldsymbol{a}, \boldsymbol{b}, \boldsymbol{c}],$$

[7] A proof of this statement can be constructed with the aid of Exercise 4(i) (p. 47).

and the required formula is obtained on cancelling the arbitrary factor $[\boldsymbol{a}, \boldsymbol{b}, \boldsymbol{c}]$.

3 THE TENSOR PRODUCT

To an ordered pair of vectors $(\boldsymbol{u}, \boldsymbol{v})$ there corresponds a tensor, denoted by $\boldsymbol{u} \otimes \boldsymbol{v}$ and called the *tensor product* of $\boldsymbol{u}$ and $\boldsymbol{v}$, which is defined through its action on an arbitrary vector $\boldsymbol{a}$ by

$$(\boldsymbol{u} \otimes \boldsymbol{v})\boldsymbol{a} = (\boldsymbol{a} \cdot \boldsymbol{v})\boldsymbol{u} \quad \forall\, \boldsymbol{a} \in \mathsf{E}. \tag{41}$$

The properties

$$\left.\begin{aligned} (\alpha\boldsymbol{u} + \beta\boldsymbol{v}) \otimes \boldsymbol{w} &= \alpha(\boldsymbol{u} \otimes \boldsymbol{w}) + \beta(\boldsymbol{v} \otimes \boldsymbol{w}) \\ \boldsymbol{u} \otimes (\alpha\boldsymbol{v} + \beta\boldsymbol{w}) &= \alpha(\boldsymbol{u} \otimes \boldsymbol{v}) + \beta(\boldsymbol{u} \otimes \boldsymbol{w}) \end{aligned}\right\} \quad \forall\, \boldsymbol{u}, \boldsymbol{v}, \boldsymbol{w} \in \mathsf{E}, \alpha, \beta \in \mathsf{R}, \tag{42}$$

and

$$(\boldsymbol{u} \otimes \boldsymbol{v})^{\mathrm{T}} = \boldsymbol{v} \otimes \boldsymbol{u} \quad \forall\, \boldsymbol{u}, \boldsymbol{v} \in \mathsf{E} \tag{43}$$

can easily be deduced from (41) by using axioms (1) and (2) and equations $(22)_1$ and (24).

Problem 7 Let $\{\boldsymbol{e}_1, \boldsymbol{e}_2, \boldsymbol{e}_3\}$ be an orthonormal basis of E. Show that

$$\boldsymbol{e}_p \otimes \boldsymbol{e}_p = \boldsymbol{I}. \tag{44}$$

Solution. Using in turn equations (41), (14) and $(23)_2$ we find that

$$(\boldsymbol{e}_p \otimes \boldsymbol{e}_p)\boldsymbol{a} = (\boldsymbol{a} \cdot \boldsymbol{e}_p)\boldsymbol{e}_p = \boldsymbol{a} = \boldsymbol{I}\boldsymbol{a} \quad \forall\, \boldsymbol{a} \in \mathsf{E}.$$

Hence the result.

Problem 8 Let $\boldsymbol{u}, \boldsymbol{v}, \boldsymbol{w}, \boldsymbol{x}$ be arbitrary vectors and $\boldsymbol{A}$ an arbitrary tensor. Show that

(i) $$(\boldsymbol{u} \otimes \boldsymbol{v})(\boldsymbol{w} \otimes \boldsymbol{x}) = (\boldsymbol{v} \cdot \boldsymbol{w})(\boldsymbol{u} \otimes \boldsymbol{x}), \tag{45}$$

(ii) $$\boldsymbol{A}(\boldsymbol{u} \otimes \boldsymbol{v}) = (\boldsymbol{A}\boldsymbol{u}) \otimes \boldsymbol{v}, \quad (\boldsymbol{u} \otimes \boldsymbol{v})\boldsymbol{A} = \boldsymbol{u} \otimes (\boldsymbol{A}^{\mathrm{T}}\boldsymbol{v}). \tag{46}$$

Solution. Let $\boldsymbol{a}$ be an arbitrary vector.

(i) $$\begin{aligned} \{(\boldsymbol{u} \otimes \boldsymbol{v})(\boldsymbol{w} \otimes \boldsymbol{x})\}\boldsymbol{a} &= (\boldsymbol{u} \otimes \boldsymbol{v})\{(\boldsymbol{w} \otimes \boldsymbol{x})\boldsymbol{a}\} = (\boldsymbol{u} \otimes \boldsymbol{v})\{(\boldsymbol{a} \cdot \boldsymbol{x})\boldsymbol{w}\} \\ &= (\boldsymbol{a} \cdot \boldsymbol{x})(\boldsymbol{u} \otimes \boldsymbol{v})\boldsymbol{w} = (\boldsymbol{a} \cdot \boldsymbol{x})(\boldsymbol{w} \cdot \boldsymbol{v})\boldsymbol{u} \\ &= (\boldsymbol{v} \cdot \boldsymbol{w})(\boldsymbol{a} \cdot \boldsymbol{x})\boldsymbol{u} = (\boldsymbol{v} \cdot \boldsymbol{w})(\boldsymbol{u} \otimes \boldsymbol{x})\boldsymbol{a}, \end{aligned}$$

where the steps are justified, in order, by equations $(22)_3$, (41), (21), (41), (1) and (41).

$$\text{(ii)}\ \{\boldsymbol{A}(\boldsymbol{u}\otimes\boldsymbol{v})\}\boldsymbol{a} = \boldsymbol{A}\{(\boldsymbol{u}\otimes\boldsymbol{v})\boldsymbol{a}\} = \boldsymbol{A}\{(\boldsymbol{a}\,.\,\boldsymbol{v})\boldsymbol{u}\}$$
$$= (\boldsymbol{a}\,.\,\boldsymbol{v})\boldsymbol{A}\boldsymbol{u} = \{(\boldsymbol{A}\boldsymbol{u})\otimes\boldsymbol{v}\}\boldsymbol{a},$$
$$\{(\boldsymbol{u}\otimes\boldsymbol{v})\boldsymbol{A}\}\boldsymbol{a} = (\boldsymbol{u}\otimes\boldsymbol{v})(\boldsymbol{A}\boldsymbol{a}) = \{(\boldsymbol{A}\boldsymbol{a})\,.\,\boldsymbol{v}\}\boldsymbol{u} = \{\boldsymbol{v}\,.\,(\boldsymbol{A}\boldsymbol{a})\}\boldsymbol{u}$$
$$= \{\boldsymbol{a}\,.\,(\boldsymbol{A}^{\mathrm{T}}\boldsymbol{v})\}\boldsymbol{u} = \{\boldsymbol{u}\otimes(\boldsymbol{A}^{\mathrm{T}}\boldsymbol{v})\}\boldsymbol{a}.$$

Here we have used the same equations as in part (i) with the addition of (24).

Problem 9 Let $\boldsymbol{u}$ and $\boldsymbol{v}$ be arbitrary vectors. Show that

$$\operatorname{tr}(\boldsymbol{u}\otimes\boldsymbol{v}) = \boldsymbol{u}\,.\,\boldsymbol{v}, \quad \det(\boldsymbol{u}\otimes\boldsymbol{v}) = 0, \tag{47}$$

and also that the second principal invariant of $\boldsymbol{u}\otimes\boldsymbol{v}$ is zero.

Solution. The required results can be derived from equations (28) to (30) by replacing $\boldsymbol{A}$ by $\boldsymbol{u}\otimes\boldsymbol{v}$ and regarding $\{\boldsymbol{a}, \boldsymbol{b}, \boldsymbol{c}\}$ as an arbitrary basis of E. On using the definition (41) we find that the second and third of the principal invariants of $\boldsymbol{u}\otimes\boldsymbol{v}$ vanish because each of the triple products on the left of (29) has two members which are scalar multiples of $\boldsymbol{u}$, while all three vectors in the triple product on the left of (30) are scalar multiples of $\boldsymbol{u}$. Equation (28) yields

$$[\boldsymbol{a}, \boldsymbol{b}, \boldsymbol{c}]\operatorname{tr}(\boldsymbol{u}\otimes\boldsymbol{v})$$
$$= [(\boldsymbol{a}\,.\,\boldsymbol{v})\boldsymbol{u}, \boldsymbol{b}, \boldsymbol{c}] + [\boldsymbol{a}, (\boldsymbol{b}\,.\,\boldsymbol{v})\boldsymbol{u}, \boldsymbol{c}] + [\boldsymbol{a}, \boldsymbol{b}, (\boldsymbol{c}\,.\,\boldsymbol{v})\boldsymbol{u}]$$
$$= (\boldsymbol{a}\,.\,\boldsymbol{v})[\boldsymbol{u}, \boldsymbol{b}, \boldsymbol{c}] + (\boldsymbol{b}\,.\,\boldsymbol{v})[\boldsymbol{a}, \boldsymbol{u}, \boldsymbol{c}] + (\boldsymbol{c}\,.\,\boldsymbol{v})[\boldsymbol{a}, \boldsymbol{b}, \boldsymbol{u}], \tag{A}$$

use being made of (10) and (11). Since $\{\boldsymbol{a}, \boldsymbol{b}, \boldsymbol{c}\}$ is a basis of E there are scalars α, β, γ such that $\boldsymbol{u} = \alpha\boldsymbol{a} + \beta\boldsymbol{b} + \gamma\boldsymbol{c}$. On entering this expression into (A) and appealing again to equations (10) and (11) we can remove the non-zero factor $[\boldsymbol{a}, \boldsymbol{b}, \boldsymbol{c}]$ from both sides, leaving

$$\operatorname{tr}(\boldsymbol{u}\otimes\boldsymbol{v}) = \alpha(\boldsymbol{a}\,.\,\boldsymbol{v}) + \beta(\boldsymbol{b}\,.\,\boldsymbol{v}) + \gamma(\boldsymbol{c}\,.\,\boldsymbol{v})$$
$$= (\alpha\boldsymbol{a} + \beta\boldsymbol{b} + \gamma\boldsymbol{c})\,.\,\boldsymbol{v} = \boldsymbol{u}\,.\,\boldsymbol{v}.$$

The elements of tensor algebra have been presented in Section 2 in an invariant form, that is without regard to a particular basis of E.

Considering now the reference of an arbitrary tensor $\boldsymbol{A}$ to the orthonormal basis $e = \{\boldsymbol{e}_1, \boldsymbol{e}_2, \boldsymbol{e}_3\}$, let A_{ij} be the components relative to e of the vector $\boldsymbol{A}\boldsymbol{e}_j$ so that

$$\boldsymbol{A}\boldsymbol{e}_j = A_{pj}\boldsymbol{e}_p \quad \text{and} \quad A_{ij} = \boldsymbol{e}_i \cdot (\boldsymbol{A}\boldsymbol{e}_j). \tag{48}$$

Selecting arbitrarily a vector $\boldsymbol{a}$ with components a_i relative to e, we find, with the aid of equations (13), (21), (41) and (12), that

$$\begin{aligned}(\boldsymbol{A} - A_{pq}\boldsymbol{e}_p \otimes \boldsymbol{e}_q)\boldsymbol{a} &= (\boldsymbol{A} - A_{pq}\boldsymbol{e}_p \otimes \boldsymbol{e}_q)(a_r\boldsymbol{e}_r)\\ &= a_r\{\boldsymbol{A}\boldsymbol{e}_r - A_{pq}(\boldsymbol{e}_r \cdot \boldsymbol{e}_q)\boldsymbol{e}_p\} = a_r(A_{pr} - A_{pq}\delta_{rq})\boldsymbol{e}_p = \boldsymbol{0}.\end{aligned}$$

Hence $\boldsymbol{A}$ admits the representation

$$\boldsymbol{A} = A_{pq}\boldsymbol{e}_p \otimes \boldsymbol{e}_q, \tag{49}$$

and the scalars A_{ij} can be appropriately referred to as the *components* of $\boldsymbol{A}$ relative to e.

An incidental result of the calculation leading to equation (49) is that the vector $\boldsymbol{A}\boldsymbol{a}$ has components $A_{ip}a_p$ relative to e, while reference to $(48)_2$ and (24) shows that the components of $\boldsymbol{A}^{\mathrm{T}}$ are A_{ji}. If $\boldsymbol{B}$ is a second arbitrary tensor with components B_{ij} relative to e, successive use of equations (49), (45) and (12) gives

$$\begin{aligned}\boldsymbol{A}\boldsymbol{B} &= (A_{pq}\boldsymbol{e}_p \otimes \boldsymbol{e}_q)(B_{rs}\boldsymbol{e}_r \otimes \boldsymbol{e}_s) = A_{pq}B_{rs}(\boldsymbol{e}_p \otimes \boldsymbol{e}_q)(\boldsymbol{e}_r \otimes \boldsymbol{e}_s)\\ &= A_{pq}B_{rs}(\boldsymbol{e}_q \cdot \boldsymbol{e}_r)(\boldsymbol{e}_p \otimes \boldsymbol{e}_s) = A_{pq}B_{rs}\delta_{qr}\boldsymbol{e}_p \otimes \boldsymbol{e}_s\\ &= A_{pq}B_{qs}\boldsymbol{e}_p \otimes \boldsymbol{e}_s.\end{aligned}$$

Thus $\boldsymbol{A}\boldsymbol{B}$ has components $A_{ip}B_{pj}$ relative to e. Again, if $\boldsymbol{u}, \boldsymbol{v}$ are arbitrary vectors with components u_i, v_i relative to e, it follows from equations (13) and (42) that the components of $\boldsymbol{u} \otimes \boldsymbol{v}$ are u_iv_j. And lastly, replacing $\{\boldsymbol{a}, \boldsymbol{b}, \boldsymbol{c}\}$ by $\{\boldsymbol{e}_1, \boldsymbol{e}_2, \boldsymbol{e}_3\}$ in equations (28) to (30) and using $(48)_1$, we obtain the following component forms of the principal invariants of $\boldsymbol{A}$:

$$I_A = \operatorname{tr} \boldsymbol{A} = A_{pp}, \tag{50}$$

$$II_A = \tfrac{1}{2}(A_{pp}^2 - A_{pq}A_{qp}), \tag{51}$$

$$III_A = \det \boldsymbol{A} = \varepsilon_{pqr}A_{p1}A_{q2}A_{r3}. \tag{52}$$

Equation (50) provides two further general properties of the trace, namely

$$\operatorname{tr} \boldsymbol{A}^{\mathrm{T}} = \operatorname{tr} \boldsymbol{A} \quad \forall\, \boldsymbol{A} \in \mathsf{L}, \qquad \operatorname{tr}(\boldsymbol{A}\boldsymbol{B}) = \operatorname{tr}(\boldsymbol{B}\boldsymbol{A}) \quad \forall\, \boldsymbol{A}, \boldsymbol{B} \in \mathsf{L}. \tag{53}$$

The reader is invited to write out the right-hand side of equation (52) as a sum of six terms and to verify that the same answer is obtained on expanding $\varepsilon_{pqr}A_{1p}A_{2q}A_{3r}$. This means that

$$\det \boldsymbol{A}^{\mathrm{T}} = \det \boldsymbol{A} \quad \forall \boldsymbol{A} \in \mathsf{L}. \tag{54}$$

4 PROPER VECTORS AND PROPER NUMBERS OF TENSORS

Let $\boldsymbol{A}$ be an arbitrary tensor. A non-zero vector $\boldsymbol{p}$ is said to be a *proper vector*[8] of $\boldsymbol{A}$ if there exists a scalar (i.e. a *real* number) λ such that

$$\boldsymbol{A}\boldsymbol{p} = \lambda\boldsymbol{p}, \quad \text{i.e.}\ (\boldsymbol{A} - \lambda\boldsymbol{I})\boldsymbol{p} = \boldsymbol{0}; \tag{55}$$

λ is called the *proper number*[8] of $\boldsymbol{A}$ associated with $\boldsymbol{p}$. Reciprocally, a scalar λ is a proper number of $\boldsymbol{A}$ if there is a non-zero vector $\boldsymbol{p}$ such that (55) holds, and in this situation $\boldsymbol{p}$ is said to be a proper vector of $\boldsymbol{A}$ associated with λ.

Problem 5 (p. 19) implies that λ is a proper number of $\boldsymbol{A}$ if and only if it is a real root of the equation

$$\det(\boldsymbol{A} - \lambda\boldsymbol{I}) = 0.$$

This is known as the *characteristic equation* of $\boldsymbol{A}$ and in view of equation (30) it can also be expressed as

$$[\boldsymbol{A}\boldsymbol{a} - \lambda\boldsymbol{a}, \boldsymbol{A}\boldsymbol{b} - \lambda\boldsymbol{b}, \boldsymbol{A}\boldsymbol{c} - \lambda\boldsymbol{c}] = 0, \tag{56}$$

where $\boldsymbol{a}, \boldsymbol{b}, \boldsymbol{c}$ are arbitrary vectors. On expanding the left-hand side of (56) with the aid of equations (10) and (11), then using the definitions (28) to (30), the arbitrary factor $[\boldsymbol{a}, \boldsymbol{b}, \boldsymbol{c}]$ can be removed and we arrive at the alternative form

$$\lambda^3 - I_{\boldsymbol{A}}\lambda^2 + II_{\boldsymbol{A}}\lambda - III_{\boldsymbol{A}} = 0 \tag{57}$$

of the characteristic equation. Since the principal invariants $I_{\boldsymbol{A}}$, $II_{\boldsymbol{A}}$, $III_{\boldsymbol{A}}$ are real, we deduce from equation (57) that $\boldsymbol{A}$ has either three proper numbers or only one.

Problem 10 Let f be a real polynomial, $\boldsymbol{A}$ an arbitrary tensor and

[8] The terms *characteristic vector*, *characteristic root* (or *value*) and *eigenvector*, *eigenvalue* are also widely used.

λ a proper number of $\boldsymbol{A}$. Show that $f(\lambda)$ is a proper number of $f(\boldsymbol{A})$ and that a proper vector of $\boldsymbol{A}$ associated with λ is also a proper vector of $f(\boldsymbol{A})$ associated with $f(\lambda)$.

Solution. Let $\boldsymbol{p}$ be a proper vector of $\boldsymbol{A}$ associated with λ. Because of equation (55) the relation

$$\boldsymbol{A}^r\boldsymbol{p} = \lambda^r\boldsymbol{p} \tag{A}$$

holds for $r = 1$. Suppose that it holds for $r = 1, 2, \ldots, n$. Then

$$\boldsymbol{A}^{n+1}\boldsymbol{p} = \boldsymbol{A}(\boldsymbol{A}^n\boldsymbol{p}) = \boldsymbol{A}(\lambda^n\boldsymbol{p}) = \lambda^n\boldsymbol{A}\boldsymbol{p} = \lambda^{n+1}\boldsymbol{p},$$

and it may be inferred, by induction, that (A) holds for all positive integers r. Since $f(\boldsymbol{A})$ is a linear combination of powers of $\boldsymbol{A}$ it follows that $f(\lambda)$ is a proper number of $f(\boldsymbol{A})$ and $\boldsymbol{p}$ an associated proper vector.

When applied to the characteristic polynomial on the left of equation (57), Problem 10 shows that if $\boldsymbol{A}$ has three proper numbers, the tensor $\boldsymbol{A}^3 - I_A\boldsymbol{A}^2 + II_A\boldsymbol{A} - III_A\boldsymbol{I}$ has three proper numbers each equal to zero. The *Cayley–Hamilton theorem*[9] asserts that, *for arbitrary* $\boldsymbol{A}$, this tensor is in fact zero;[10] that, in other words, a tensor satisfies its own characteristic equation:

$$\boldsymbol{A}^3 - I_A\boldsymbol{A}^2 + II_A\boldsymbol{A} - III_A\boldsymbol{I} = \boldsymbol{O} \quad \forall\, \boldsymbol{A} \in \mathsf{L}. \tag{58}$$

5 SYMMETRIC TENSORS

A symmetric tensor $\boldsymbol{S}$ possesses three proper numbers ($\lambda_1, \lambda_2, \lambda_3$, say) and an orthonormal set of proper vectors, $\boldsymbol{p}_1, \boldsymbol{p}_2, \boldsymbol{p}_3$, associated respectively with $\lambda_1, \lambda_2, \lambda_3$.[11] Using successively equations (44), $(46)_1$, (55) and $(42)_1$, we can express $\boldsymbol{S}$ in terms of λ_i and $\boldsymbol{p}_i$ ($i = 1, 2, 3$) as follows:

$$\boldsymbol{S} = \boldsymbol{S}\boldsymbol{I} = \boldsymbol{S}(\boldsymbol{p}_r \otimes \boldsymbol{p}_r) = (\boldsymbol{S}\boldsymbol{p}_r) \otimes \boldsymbol{p}_r = \sum_{r=1}^{3} \lambda_r(\boldsymbol{p}_r \otimes \boldsymbol{p}_r). \tag{59}$$

[9] See K. Hoffman and R. Kunze, *op. cit.* p. 166.

[10] Problem 9 (p. 22) shows that if $\boldsymbol{u}$ and $\boldsymbol{v}$ are non-zero orthogonal vectors, all the principal invariants of $\boldsymbol{u} \otimes \boldsymbol{v}$ are zero. This furnishes an example of a non-zero tensor possessing three proper numbers all equal to zero.

[11] See K. Hoffman and R. Kunze, *op. cit.* p. 264.

This result is referred to as the *spectral representation* of a symmetric tensor. With the existence of three proper numbers assured, the expressions

$$I_S = \operatorname{tr} \boldsymbol{S} = \lambda_1 + \lambda_2 + \lambda_3, \tag{60}$$

$$II_S = \lambda_2\lambda_3 + \lambda_3\lambda_1 + \lambda_1\lambda_2, \tag{61}$$

$$III_S = \det \boldsymbol{S} = \lambda_1\lambda_2\lambda_3, \tag{62}$$

for the principal invariants of $\boldsymbol{S}$ follow directly from equation (57).

A tensor $\boldsymbol{A}$ is said to be *positive semi-definite* if

$$\boldsymbol{a} \cdot (\boldsymbol{A}\boldsymbol{a}) \geqslant 0 \quad \forall\, \boldsymbol{a} \in \mathsf{E}. \tag{63}$$

If the stronger condition

$$\boldsymbol{a} \cdot (\boldsymbol{A}\boldsymbol{a}) > 0 \quad \forall \text{ non-zero } \boldsymbol{a} \in \mathsf{E} \tag{64}$$

is fulfilled, $\boldsymbol{A}$ is said to be *positive definite*. The terms *negative semi-definite* and *negative definite* apply to tensors whose negatives are positive semi-definite and positive definite respectively.

Let $\boldsymbol{S}$ be a positive semi-definite symmetric tensor with proper numbers λ_i and associated unit proper vectors $\boldsymbol{p}_i$ $(i = 1, 2, 3)$. Then the definition (63), in conjunction with equations (59) and (41), gives

$$\boldsymbol{a} \cdot (\boldsymbol{S}\boldsymbol{a}) = \lambda_r(\boldsymbol{a} \cdot \boldsymbol{p}_r)^2 \geqslant 0 \quad \forall\, \boldsymbol{a} \in \mathsf{E},$$

and it follows that $\lambda_1 \geqslant 0, \lambda_2 \geqslant 0, \lambda_3 \geqslant 0$. The tensor $\boldsymbol{S}^{1/2}$ defined by

$$\boldsymbol{S}^{1/2} = \sum_{r=1}^{3} \lambda_r^{1/2}(\boldsymbol{p}_r \otimes \boldsymbol{p}_r) \tag{65}$$

is clearly symmetric and positive semi-definite and, in consequence of equations (42), (45) and (59), its square is $\boldsymbol{S}$. If $\bar{\boldsymbol{S}}^{1/2}$ is a second tensor with these properties, Problem 10 (p. 24) shows that a proper vector of $\bar{\boldsymbol{S}}^{1/2}$ is also a proper vector of $\boldsymbol{S}$ and that the square of the associated proper number of $\bar{\boldsymbol{S}}^{1/2}$ is a proper number of $\boldsymbol{S}$. Thus $\bar{\boldsymbol{S}}^{1/2} = \boldsymbol{S}^{1/2}$ and we conclude that a positive semi-definite symmetric tensor has a unique positive semi-definite symmetric square root with the spectral representation (65).

A positive definite symmetric tensor $\boldsymbol{S}$ has positive proper numbers. Hence, from equation (62), $\det \boldsymbol{S} > 0$ implying that $\boldsymbol{S}$ is invertible.

The spectral representation of the inverse is

$$S^{-1} = \sum_{r=1}^{3} \lambda_r^{-1}(\boldsymbol{p}_r \otimes \boldsymbol{p}_r). \tag{66}$$

Problem 11 Let A be an arbitrary tensor. Show that $A^{\mathrm{T}}A$ and AA^{T} are positive semi-definite symmetric tensors. If A is invertible, prove that these tensors are positive definite.

Solution. From equation (26),

$$(A^{\mathrm{T}}A)^{\mathrm{T}} = A^{\mathrm{T}}(A^{\mathrm{T}})^{\mathrm{T}} = A^{\mathrm{T}}A, \quad (AA^{\mathrm{T}})^{\mathrm{T}} = (A^{\mathrm{T}})^{\mathrm{T}}A^{\mathrm{T}} = AA^{\mathrm{T}}.$$

This confirms that $A^{\mathrm{T}}A$ and AA^{T} are symmetric. Let $\boldsymbol{a}$ be an arbitrary vector. Then, with the use of equation (24),

$$\left.\begin{aligned} \boldsymbol{a} \cdot \{(A^{\mathrm{T}}A)\boldsymbol{a}\} &= \boldsymbol{a} \cdot \{A^{\mathrm{T}}(A\boldsymbol{a})\} = (A\boldsymbol{a}) \cdot (A\boldsymbol{a}), \\ \boldsymbol{a} \cdot \{(AA^{\mathrm{T}})\boldsymbol{a}\} &= \qquad\qquad\qquad\quad (A^{\mathrm{T}}\boldsymbol{a}) \cdot (A^{\mathrm{T}}\boldsymbol{a}). \end{aligned}\right\} \tag{A}$$

Axiom (3) guarantees that the scalar products on the right-hand sides of (A) are non-negative whence, by the definition (63), $A^{\mathrm{T}}A$ and AA^{T} are positive semi-definite tensors. If A is invertible, it follows from Problem 5 (p. 19) that $A\boldsymbol{a}$ and $A^{\mathrm{T}}\boldsymbol{a}$ are non-zero whenever $\boldsymbol{a} \neq \boldsymbol{0}$. In this case, therefore, with further appeal to axiom (3), the final scalar products in (A) are positive for all non-zero $\boldsymbol{a}$ and it follows from (64) that $A^{\mathrm{T}}A$ and AA^{T} are positive definite.

Problem 12 Given a non-zero symmetric tensor T, show that an arbitrary symmetric tensor S can be expressed uniquely in the form $S = \alpha T + U$ where α is a scalar and U a symmetric tensor such that $\mathrm{tr}(TU) = 0$. Hence prove that if T_1 and T_2 are non-zero symmetric tensors and $\mathrm{tr}(T_1A) = 0$ for all tensors A such that $\mathrm{tr}(T_2A) = 0$, then T_1 is a scalar multiple of T_2.

Solution. Since T is non-zero, reference to the spectral representation (59) shows that at least one of its proper numbers is non-zero. From Problem 10 (p. 24), the proper numbers of T^2 are the squares of the proper numbers of T. It is therefore a consequence of equation (60) that $\mathrm{tr}\, T^2 > 0$. Evidently a symmetric tensor whose square has zero trace is necessarily the zero tensor.

Corresponding to the arbitrary symmetric tensor $\boldsymbol{S}$ we now define

$$\alpha = \text{tr}(\boldsymbol{TS})/\text{tr}\, \boldsymbol{T}^2, \quad \boldsymbol{U} = \boldsymbol{S} - \alpha \boldsymbol{T}.$$

Then

$$\boldsymbol{S} = \alpha\boldsymbol{T} + \boldsymbol{U} \quad \text{and} \quad \text{tr}(\boldsymbol{TU}) = \text{tr}(\boldsymbol{TS}) - \alpha\, \text{tr}\, \boldsymbol{T}^2 = 0,$$

which demonstrates that $\boldsymbol{S}$ can be given the stated representation. To prove that it is unique, suppose that $\boldsymbol{S} = \beta\boldsymbol{T} + \boldsymbol{V}$ where $\text{tr}(\boldsymbol{TV}) = 0$. Then

$$(\beta - \alpha)\boldsymbol{T} = \boldsymbol{U} - \boldsymbol{V} \quad \text{and} \quad (\beta - \alpha)\, \text{tr}\, \boldsymbol{T}^2 = \text{tr}(\boldsymbol{TU}) - \text{tr}(\boldsymbol{TV}) = 0.$$

Since $\text{tr}\, \boldsymbol{T}^2 > 0$ it follows that $\beta = \alpha$ and $\boldsymbol{V} = \boldsymbol{U}$.

The result just established enables us to write $\boldsymbol{T}_1 = \alpha\boldsymbol{T}_2 + \boldsymbol{U}$ where $\text{tr}(\boldsymbol{T}_2\boldsymbol{U}) = 0$. Since $\text{tr}(\boldsymbol{T}_1\boldsymbol{A}) = 0$ whenever $\text{tr}(\boldsymbol{T}_2\boldsymbol{A}) = 0$, $\boldsymbol{U}$ must satisfy the condition $\text{tr}(\boldsymbol{UA}) = 0$ for all tensors $\boldsymbol{A}$ such that $\text{tr}(\boldsymbol{T}_2\boldsymbol{A}) = 0$. But $\text{tr}(\boldsymbol{T}_2\boldsymbol{U}) = 0$. Hence $\text{tr}\, \boldsymbol{U}^2 = 0$ from which we deduce that $\boldsymbol{U} = \mathbf{O}$ and $\boldsymbol{T}_1 = \alpha\boldsymbol{T}_2$.

6 SKEW-SYMMETRIC TENSORS

We turn next to the main properties of skew-symmetric tensors. Let $\boldsymbol{W}$ be a tensor of this type (so that $\boldsymbol{W}^{\text{T}} = -\boldsymbol{W}$) and let $\boldsymbol{a}$, $\boldsymbol{b}$ be arbitrary vectors. From equation (24),

$$\boldsymbol{b} \,.\, (\boldsymbol{Wa}) = \boldsymbol{a} \,.\, (\boldsymbol{W}^{\text{T}}\boldsymbol{b}) = -\boldsymbol{a} \,.\, (\boldsymbol{Wb}), \tag{67}$$

and on setting $\boldsymbol{b} = \boldsymbol{a}$ we have

$$\boldsymbol{a} \,.\, (\boldsymbol{Wa}) = 0. \tag{68}$$

These results, taken in conjunction with equation $(48)_2$, show that, relative to an orthonormal basis of E, three of the components of $\boldsymbol{W}$ vanish and three others are equal to the negatives of the remaining three. Thus a skew-symmetric tensor has only three independent components, which suggests that, in some way, it is equivalent to a vector. The discussion which follows elicits the nature of this relationship.

A skew-symmetric tensor $\boldsymbol{W}$, like any tensor, is known to possess at least one proper number, λ say. Let $\boldsymbol{p}$ be one of the two associated proper vectors with unit norm. Then $\boldsymbol{Wp} = \lambda\boldsymbol{p}$ and on forming the scalar product of each side with $\boldsymbol{p}$, $\lambda = \boldsymbol{p} . (\boldsymbol{Wp}) = 0$, in view of (68).

This means that $\boldsymbol{W}$ has either a single proper number which is zero or three proper numbers each equal to zero. It also establishes the existence of a unit vector $\boldsymbol{p}$ such that $\boldsymbol{Wp} = \boldsymbol{0}$.

Let $\boldsymbol{q}$ and $\boldsymbol{r}$ be unit vectors forming with $\boldsymbol{p}$ an orthonormal basis of E which is positive in E^+. Then

$$\boldsymbol{p} = \boldsymbol{q} \wedge \boldsymbol{r}, \quad \boldsymbol{q} = \boldsymbol{r} \wedge \boldsymbol{p}, \quad \boldsymbol{r} = \boldsymbol{p} \wedge \boldsymbol{q}, \quad [\boldsymbol{p}, \boldsymbol{q}, \boldsymbol{r}] = 1. \tag{69}$$

Referred to the basis $\{\boldsymbol{p}, \boldsymbol{q}, \boldsymbol{r}\}$, via equations $(48)_2$, (49), (67) and (68), $\boldsymbol{W}$ takes the simple form

$$\boldsymbol{W} = \omega(\boldsymbol{r} \otimes \boldsymbol{q} - \boldsymbol{q} \otimes \boldsymbol{r}), \tag{70}$$

where $\omega = \boldsymbol{r} \, . \, (\boldsymbol{Wq})$ must be non-zero if $\boldsymbol{W}$ is not merely the zero tensor. Now set $\boldsymbol{w} = \omega \boldsymbol{p}$ and let $\boldsymbol{a}$ be an arbitrary vector. Then

$$\begin{aligned} \boldsymbol{Wa} - \boldsymbol{w} \wedge \boldsymbol{a} &= \omega[(\boldsymbol{r} \otimes \boldsymbol{q})\boldsymbol{a} - (\boldsymbol{q} \otimes \boldsymbol{r})\boldsymbol{a} \\ &\qquad - \boldsymbol{p} \wedge \{(\boldsymbol{a} \, . \, \boldsymbol{p})\boldsymbol{p} + (\boldsymbol{a} \, . \, \boldsymbol{q})\boldsymbol{q} + (\boldsymbol{a} \, . \, \boldsymbol{r})\boldsymbol{r}\}] \\ &= \omega\{(\boldsymbol{a} \, . \, \boldsymbol{q})(\boldsymbol{r} - \boldsymbol{p} \wedge \boldsymbol{q}) - (\boldsymbol{a} \, . \, \boldsymbol{r})(\boldsymbol{q} + \boldsymbol{p} \wedge \boldsymbol{r})\} = \boldsymbol{0}, \end{aligned}$$

use being made of equations (70), (14), (41) and (69). We have thus proved that, associated with a skew-symmetric tensor $\boldsymbol{W}$, there is a vector $\boldsymbol{w}$ such that

$$\boldsymbol{Wa} = \boldsymbol{w} \wedge \boldsymbol{a} \quad \forall\, \boldsymbol{a} \in \mathsf{E}: \tag{71}$$

$\boldsymbol{w}$ is called the *axial vector* of $\boldsymbol{W}$.

Problem 13 Let $\boldsymbol{W}$ be an arbitrary skew-symmetric tensor and $\boldsymbol{w}$ its axial vector. Show that

$$I_W = \operatorname{tr} \boldsymbol{W} = 0, \quad II_W = |\boldsymbol{w}|^2, \quad III_W = \det \boldsymbol{W} = 0. \tag{72}$$

Deduce that $\boldsymbol{W}$ has only one proper number.

Solution. We use equations (28) to (30) to evaluate the principal invariants of $\boldsymbol{W}$, $\boldsymbol{A}$ being replaced by $\boldsymbol{W}$ and $\{\boldsymbol{a}, \boldsymbol{b}, \boldsymbol{c}\}$ by $\{\boldsymbol{p}, \boldsymbol{q}, \boldsymbol{r}\}$, the orthonormal basis introduced above. Since $\boldsymbol{Wp} = \boldsymbol{0}$ and $[\boldsymbol{p}, \boldsymbol{q}, \boldsymbol{r}] = 1$ we obtain

$$I_W = [\boldsymbol{p}, \boldsymbol{Wq}, \boldsymbol{r}] + [\boldsymbol{p}, \boldsymbol{q}, \boldsymbol{Wr}], \quad II_W = [\boldsymbol{p}, \boldsymbol{Wq}, \boldsymbol{Wr}], \quad III_W = 0.$$

From equations (71) and (69), bearing in mind that $\boldsymbol{w} = \omega \boldsymbol{p}$,

$$\boldsymbol{Wq} = \omega \boldsymbol{p} \wedge \boldsymbol{q} = \omega \boldsymbol{r}, \quad \boldsymbol{Wr} = \omega \boldsymbol{p} \wedge \boldsymbol{r} = -\omega \boldsymbol{q}.$$

Hence $I_W = 0$ and

$$II_W = [\boldsymbol{p}, \omega \boldsymbol{r}, -\omega \boldsymbol{q}] = \omega^2[\boldsymbol{p}, \boldsymbol{r}, -\boldsymbol{q}] = \omega^2 = |\boldsymbol{w}|^2.$$

The characteristic equation of $\boldsymbol{W}$, derived by substituting from (72) into (57), is $\lambda^3 + |\boldsymbol{w}|^2\lambda = 0$. This has a single real root, equal to zero.

Problem 14 Let $\boldsymbol{u}$ and $\boldsymbol{v}$ be arbitrary vectors. Show that $\boldsymbol{v} \otimes \boldsymbol{u} - \boldsymbol{u} \otimes \boldsymbol{v}$ is a skew-symmetric tensor and that $\boldsymbol{u} \wedge \boldsymbol{v}$ is its axial vector.

Solution. The skew-symmetric property of $\boldsymbol{v} \otimes \boldsymbol{u} - \boldsymbol{u} \otimes \boldsymbol{v}$ is verified with the aid of equations (25) and (43):

$$\begin{aligned}(\boldsymbol{v} \otimes \boldsymbol{u} - \boldsymbol{u} \otimes \boldsymbol{v})^{\mathrm{T}} &= (\boldsymbol{v} \otimes \boldsymbol{u})^{\mathrm{T}} - (\boldsymbol{u} \otimes \boldsymbol{v})^{\mathrm{T}} \\ &= \boldsymbol{u} \otimes \boldsymbol{v} - \boldsymbol{v} \otimes \boldsymbol{u} = -(\boldsymbol{v} \otimes \boldsymbol{u} - \boldsymbol{u} \otimes \boldsymbol{v}).\end{aligned}$$

We observe next that

$$(\boldsymbol{v} \otimes \boldsymbol{u} - \boldsymbol{u} \otimes \boldsymbol{v})(\boldsymbol{u} \wedge \boldsymbol{v}) = \{\boldsymbol{u} \,.\, (\boldsymbol{u} \wedge \boldsymbol{v})\}\boldsymbol{v} - \{\boldsymbol{v} \,.\, (\boldsymbol{u} \wedge \boldsymbol{v})\}\boldsymbol{u} = \boldsymbol{0},$$

whence, recalling the reasoning leading to equation (71), the axial vector of $\boldsymbol{v} \otimes \boldsymbol{u} - \boldsymbol{u} \otimes \boldsymbol{v}$ is a scalar multiple of $\boldsymbol{u} \wedge \boldsymbol{v}$. Accordingly,

$$(\boldsymbol{v} \otimes \boldsymbol{u} - \boldsymbol{u} \otimes \boldsymbol{v})\boldsymbol{a} = (\boldsymbol{a} \,.\, \boldsymbol{u})\boldsymbol{v} - (\boldsymbol{a} \,.\, \boldsymbol{v})\boldsymbol{u} = \alpha(\boldsymbol{u} \wedge \boldsymbol{v}) \wedge \boldsymbol{a} \qquad \forall\, \boldsymbol{a} \in \mathsf{E}, \qquad \text{(A)}$$

and to complete the solution we must show that $\alpha = 1$. Setting $\boldsymbol{a} = \boldsymbol{u}$ in (A) and then forming the scalar product of each side with $\boldsymbol{v}$ we have

$$(\boldsymbol{u} \,.\, \boldsymbol{u})(\boldsymbol{v} \,.\, \boldsymbol{v}) - (\boldsymbol{u} \,.\, \boldsymbol{v})^2 = \alpha\{(\boldsymbol{u} \wedge \boldsymbol{v}) \wedge \boldsymbol{u}\} \,.\, \boldsymbol{v} = \alpha(\boldsymbol{u} \wedge \boldsymbol{v}) \,.\, (\boldsymbol{u} \wedge \boldsymbol{v}).$$

Thus, in view of axiom (7), $\alpha = 1$.

[The useful identities

$$\boldsymbol{a} \wedge (\boldsymbol{b} \wedge \boldsymbol{c}) = (\boldsymbol{b} \otimes \boldsymbol{c} - \boldsymbol{c} \otimes \boldsymbol{b})\boldsymbol{a} = (\boldsymbol{a} \,.\, \boldsymbol{c})\boldsymbol{b} - (\boldsymbol{a} \,.\, \boldsymbol{b})\boldsymbol{c} \qquad \forall\, \boldsymbol{a}, \boldsymbol{b}, \boldsymbol{c} \in \mathsf{E} \qquad (73)$$

which now follow from (A) are noted for future reference.]

7 ORTHOGONAL TENSORS

The third special class of tensors considered in this chapter has as its defining property the preservation of scalar products: a tensor $\boldsymbol{Q}$

such that

$$(Qa).(Qb) = a.b \quad \forall\, a, b \in \mathsf{E} \tag{74}$$

is called an *orthogonal tensor.* Since

$$(Qa).(Qb) = b.\{Q^{\mathrm{T}}(Qa)\} = b.\{(Q^{\mathrm{T}}Q)a\},$$

a necessary and sufficient condition for Q to be orthogonal is

$$Q^{\mathrm{T}}Q = I. \tag{75}$$

Equations (33), (54) and (34) give $\det(Q^{\mathrm{T}}Q) = (\det Q)^2 = 1$. Thus Q is invertible and, from (35) and (75),

$$Q^{-1} = Q^{\mathrm{T}}, \quad QQ^{\mathrm{T}} = I. \tag{76}$$

Q is said to be a *proper orthogonal tensor* if $\det Q = 1$ and an *improper orthogonal tensor* if $\det Q = -1$. By virtue of equation (32), the negative of an improper orthogonal tensor is proper orthogonal. In the remainder of this section we assume that $\det Q = 1$.

The identity $Q^{\mathrm{T}}(Q - I) = -(Q - I)^{\mathrm{T}}$ is a direct consequence of equation (75). On forming the determinant of each side and using equations (33), (54) and (32) we find that $\det(Q - I) = 0$, indicating that Q has a proper number equal to 1. Accordingly there exists a unit vector p such that

$$Qp = p = Q^{\mathrm{T}}p. \tag{77}$$

If q and r are unit vectors, related to p in the same way as in Section 6, we deduce from (77) that

$$q.(Qp) = r.(Qp) = p.(Qq) = p.(Qr) = 0, \; p.(Qp) = 1, \tag{78}$$

and from (74) that

$$(Qq).(Qr) = 0, \quad |Qq| = |Qr| = 1.$$

In particular, q, r and Qq, Qr are orthogonal pairs of unit vectors all orthogonal to p and therefore satisfying relations of the form

$$Qq = \alpha q + \beta r, \quad Qr = \gamma q + \delta r,$$

where $\alpha^2 + \beta^2 = 1$, $\gamma^2 + \delta^2 = 1$, $\alpha\gamma + \beta\delta = 0$. Further, from equations (30), $(69)_4$ and $(77)_1$,

$$\det Q = [Qp, Qq, Qr] = [p, \alpha q + \beta r, \gamma q + \delta r] = \alpha\delta - \beta\gamma = 1.$$

The four connections between the scalars $\alpha, \beta, \gamma, \delta$ assure the existence of an angle $\theta(-\pi<\theta\leqslant\pi)$ such that $\alpha=\delta=\cos\theta$, $\beta=-\gamma=\sin\theta$, and the set of scalar products (78) is completed by

$$-\boldsymbol{q}\cdot(\boldsymbol{Q}\boldsymbol{r}) = \boldsymbol{r}\cdot(\boldsymbol{Q}\boldsymbol{q}) = \sin\theta,\ \boldsymbol{q}\cdot(\boldsymbol{Q}\boldsymbol{q}) = \boldsymbol{r}\cdot(\boldsymbol{Q}\boldsymbol{r}) = \cos\theta.$$

The result of referring $\boldsymbol{Q}$ to the orthonormal basis $\{\boldsymbol{p}, \boldsymbol{q}, \boldsymbol{r}\}$, via equations $(48)_2$ and (49), is therefore

$$\boldsymbol{Q} = \boldsymbol{p}\otimes\boldsymbol{p} + (\boldsymbol{q}\otimes\boldsymbol{q} + \boldsymbol{r}\otimes\boldsymbol{r})\cos\theta - (\boldsymbol{q}\otimes\boldsymbol{r} - \boldsymbol{r}\otimes\boldsymbol{q})\sin\theta. \quad (79)$$

This representation is shown in Section 9 to allow a natural interpretation of an arbitrary proper orthogonal tensor as a rotation.

Problem 15 Let $\boldsymbol{Q}$ be a proper orthogonal tensor. Show that, in terms of the angle θ appearing in the representation (79),

$$I_{\boldsymbol{Q}} = II_{\boldsymbol{Q}} = 1 + 2\cos\theta.$$

Deduce that if $\theta \neq 0, \pi$, $\boldsymbol{Q}$ has only one proper number.

Solution. Proceeding as in Problem 13 (p. 29) we deduce from equations (28) and (29), with the aid of $(69)_4$ and $(77)_1$, the results

$$I_{\boldsymbol{Q}} = 1 + [\boldsymbol{p}, \boldsymbol{Q}\boldsymbol{q}, \boldsymbol{r}] + [\boldsymbol{p}, \boldsymbol{q}, \boldsymbol{Q}\boldsymbol{r}],$$
$$II_{\boldsymbol{Q}} = [\boldsymbol{Q}\boldsymbol{p}, \boldsymbol{Q}\boldsymbol{q}, \boldsymbol{Q}\boldsymbol{r}] + [\boldsymbol{p}, \boldsymbol{q}, \boldsymbol{Q}\boldsymbol{r}] + [\boldsymbol{p}, \boldsymbol{Q}\boldsymbol{q}, \boldsymbol{r}].$$

Also, from (30),

$$[\boldsymbol{Q}\boldsymbol{p}, \boldsymbol{Q}\boldsymbol{q}, \boldsymbol{Q}\boldsymbol{r}] = \det\boldsymbol{Q} = 1,$$

and, as shown above,

$$\boldsymbol{Q}\boldsymbol{q} = \cos\theta\boldsymbol{q} + \sin\theta\boldsymbol{r}, \quad \boldsymbol{Q}\boldsymbol{r} = -\sin\theta\boldsymbol{q} + \cos\theta\boldsymbol{r}.$$

Thus

$$I_{\boldsymbol{Q}} = II_{\boldsymbol{Q}} = 1 + [\boldsymbol{p}, \cos\theta\boldsymbol{q} + \sin\theta\boldsymbol{r}, \boldsymbol{r}] + [\boldsymbol{p}, \boldsymbol{q}, -\sin\theta\boldsymbol{q} + \cos\theta\boldsymbol{r}]$$
$$= 1 + 2\cos\theta[\boldsymbol{p}, \boldsymbol{q}, \boldsymbol{r}] = 1 + 2\cos\theta.$$

The characteristic equation of $\boldsymbol{Q}$ can now be written down, by making the appropriate substitutions in (57), and factorized as

$$(\lambda - 1)(\lambda^2 - 2\lambda\cos\theta + 1) = 0.$$

If $\theta \neq 0, \pi$, $|\cos\theta| < 1$ and the quadratic factor has a negative discriminant. Under these restrictions Q therefore has a single proper number, equal to 1.

8 POLAR DECOMPOSITIONS

It has been pointed out in Section 2 that an arbitrary tensor has a unique additive decomposition into symmetric and skew-symmetric parts (equation (27)). Of comparable significance in continuum mechanics are the multiplicative decompositions afforded by the *polar decomposition theorem*. This result states that *an arbitrary invertible tensor* A *can be expressed in the forms*

$$A = QU = VQ, \tag{80}$$

where Q *is an orthogonal tensor and* U, V *are positive definite symmetric tensors. Moreover the right polar decomposition* $(80)_1$ *and the left polar decomposition* $(80)_2$ *are unique.*[12]

An elementary proof of the theorem starts from the observation that since $A^{\mathrm{T}}A$ and AA^{T} are positive definite symmetric tensors (by Problem 11, p. 27), they have unique positive definite symmetric square roots, U and V respectively. The tensors $Q = AU^{-1}$ and $R = V^{-1}A$ are orthogonal: for

$$\begin{aligned} Q^{\mathrm{T}}Q &= (AU^{-1})^{\mathrm{T}}(AU^{-1}) = (U^{-1}A^{\mathrm{T}})(AU^{-1}) = U^{-1}(A^{\mathrm{T}}A)U^{-1} \\ &= U^{-1}U^2U^{-1} = I, \\ RR^{\mathrm{T}} &= (V^{-1}A)(V^{-1}A)^{\mathrm{T}} = (V^{-1}A)(A^{\mathrm{T}}V^{-1}) = V^{-1}(AA^{\mathrm{T}})V^{-1} \\ &= V^{-1}V^2V^{-1} = I. \end{aligned}$$

Thus

$$A = QU = VR.$$

Suppose that A has a second right polar decomposition $A = \overline{Q}\,\overline{U}$ where $\overline{Q}$ is orthogonal and $\overline{U}$ positive definite and symmetric. Then

$$A^{\mathrm{T}}A = (\overline{Q}\,\overline{U})^{\mathrm{T}}(\overline{Q}\,\overline{U}) = (\overline{U}\,\overline{Q}^{\mathrm{T}})(\overline{Q}\,\overline{U}) = \overline{U}(\overline{Q}^{\mathrm{T}}\overline{Q})\overline{U} = \overline{U}^2,$$

[12] There is a noteworthy parallel between this theorem and the polar representation $z = re^{i\theta}$ of a complex number z. The positive definite tensors in (80) correspond to the modulus r and the orthogonal tensor to the factor $e^{i\theta}$ which can be interpreted as a rotation of the complex plane.

from which it follows that $\overline{\boldsymbol{U}} = \boldsymbol{U}$, since $\boldsymbol{A}^{\mathrm{T}}\boldsymbol{A}$ has only one positive definite symmetric square root, and

$$\overline{\boldsymbol{Q}} = \boldsymbol{A}\overline{\boldsymbol{U}}^{-1} = \boldsymbol{A}\boldsymbol{U}^{-1} = \boldsymbol{Q}.$$

A similar argument establishes the uniqueness of the left polar decomposition $\boldsymbol{A} = \boldsymbol{V}\boldsymbol{R}$.

It remains only to show that $\boldsymbol{R} = \boldsymbol{Q}$. In view of the orthogonality of $\boldsymbol{R}$ and the possession by $\boldsymbol{V}$ of a positive definite square root, we have

$$\boldsymbol{A} = \boldsymbol{Q}\boldsymbol{U} = (\boldsymbol{R}\boldsymbol{R}^{\mathrm{T}})(\boldsymbol{V}\boldsymbol{R}) = \boldsymbol{R}(\boldsymbol{R}^{\mathrm{T}}\boldsymbol{V}\boldsymbol{R}) = \boldsymbol{R}\{(\boldsymbol{V}^{1/2}\boldsymbol{R})^{\mathrm{T}}(\boldsymbol{V}^{1/2}\boldsymbol{R})\},$$

from which $\boldsymbol{Q}\boldsymbol{U}$ and $\boldsymbol{R}(\boldsymbol{R}^{\mathrm{T}}\boldsymbol{V}\boldsymbol{R})$ are recognized as alternative right polar decompositions of $\boldsymbol{A}$. Hence, by the uniqueness result already proved, $\boldsymbol{Q} = \boldsymbol{R}$ and $\boldsymbol{U} = \boldsymbol{R}^{\mathrm{T}}\boldsymbol{V}\boldsymbol{R}$.

It should be noted that $\boldsymbol{Q}$ is proper or improper orthogonal according as $\det \boldsymbol{A}$ is positive or negative.

Problem 16 Let $\boldsymbol{A}$ be an invertible tensor with right and left polar decompositions $\boldsymbol{Q}\boldsymbol{U}$ and $\boldsymbol{V}\boldsymbol{Q}$. Since $\boldsymbol{U}$ and $\boldsymbol{V}$ are symmetric tensors they each possess three proper numbers and an orthonormal set of associated proper vectors: let λ_i and $\boldsymbol{p}_i$ ($i = 1, 2, 3$) be the proper numbers and vectors of $\boldsymbol{U}$ and μ_i, $\boldsymbol{q}_i$ the corresponding quantities for $\boldsymbol{V}$. Show that

$$\mu_i = \lambda_i, \quad \boldsymbol{q}_i = \boldsymbol{Q}\boldsymbol{p}_i, \tag{81}$$

and obtain representations of $\boldsymbol{U}$, $\boldsymbol{V}$, $\boldsymbol{Q}$, $\boldsymbol{A}$ and $\boldsymbol{A}^{-1}$ in terms of λ_i, $\boldsymbol{p}_i$ and $\boldsymbol{q}_i$.

Solution. From equation (55) the proper numbers and vectors of $\boldsymbol{U}$ are connected by $\boldsymbol{U}\boldsymbol{p}_i = \lambda_i\boldsymbol{p}_i$. Also $\boldsymbol{Q}\boldsymbol{U} = \boldsymbol{V}\boldsymbol{Q}$ $(=\boldsymbol{A})$. Hence

$$\boldsymbol{V}(\boldsymbol{Q}\boldsymbol{p}_i) = (\boldsymbol{V}\boldsymbol{Q})\boldsymbol{p}_i = (\boldsymbol{Q}\boldsymbol{U})\boldsymbol{p}_i = \boldsymbol{Q}(\boldsymbol{U}\boldsymbol{p}_i) = \boldsymbol{Q}(\lambda_i\boldsymbol{p}_i) = \lambda_i\boldsymbol{Q}\boldsymbol{p}_i,$$

which shows that λ_i is a proper number of $\boldsymbol{V}$ and $\boldsymbol{Q}\boldsymbol{p}_i$ an associated proper vector. It follows that the λ_i, suitably ordered, are equal to the μ_i and that the unit vectors $\boldsymbol{Q}\boldsymbol{p}_i$, appropriately ordered and signed, are equal to the $\boldsymbol{q}_i$.

Expressions for $\boldsymbol{U}$ and $\boldsymbol{V}$ in terms of λ_i, $\boldsymbol{p}_i$ and $\boldsymbol{q}_i$ are provided by the spectral representations

$$U = \sum_{r=1}^{3} \lambda_r(\boldsymbol{p}_r \otimes \boldsymbol{p}_r), \quad V = \sum_{r=1}^{3} \lambda_r(\boldsymbol{q}_r \otimes \boldsymbol{q}_r). \tag{82}$$

With the aid of equations (44), (46) and $(81)_2$ we find that

$$Q = QI = Q(\boldsymbol{p}_r \otimes \boldsymbol{p}_r) = (Q\boldsymbol{p}_r) \otimes \boldsymbol{p}_r = \boldsymbol{q}_r \otimes \boldsymbol{p}_r, \tag{83}$$

while similar manipulations, involving also equations $(76)_1$ and (66), give

$$\begin{aligned} A = QU &= Q \sum_{r=1}^{3} \lambda_r(\boldsymbol{p}_r \otimes \boldsymbol{p}_r) = \sum_{r=1}^{3} \lambda_r Q(\boldsymbol{p}_r \otimes \boldsymbol{p}_r) \\ &= \sum_{r=1}^{3} \lambda_r\{(Q\boldsymbol{p}_r) \otimes \boldsymbol{p}_r\} = \sum_{r=1}^{3} \lambda_r(\boldsymbol{q}_r \otimes \boldsymbol{p}_r), \end{aligned} \tag{84}$$

$$\begin{aligned} A^{-1} = (QU)^{-1} = U^{-1}Q^{\mathrm{T}} &= \left(\sum_{r=1}^{3} \lambda_r^{-1}(\boldsymbol{p}_r \otimes \boldsymbol{p}_r)\right) Q^{\mathrm{T}} \\ &= \sum_{r=1}^{3} \lambda_r^{-1}\{(\boldsymbol{p}_r \otimes \boldsymbol{p}_r)Q^{\mathrm{T}}\} = \sum_{r=1}^{3} \lambda_r^{-1}\{\boldsymbol{p}_r \otimes (Q\boldsymbol{p}_r)\} \\ &= \sum_{r=1}^{3} \lambda_r^{-1}(\boldsymbol{p}_r \otimes \boldsymbol{q}_r). \end{aligned} \tag{85}$$

9 GEOMETRICAL CONSIDERATIONS: COORDINATES

The term *Euclidean point space* applies to a set $\mathfrak{E}$, with elements called *points*, which is related to a Euclidean vector space E in the following manner. To each ordered pair of points (x, y) in $\mathfrak{E}$ there corresponds a unique vector in E, denoted by $\vec{xy}$, with the properties

(i) $\vec{yx} = -\vec{xy} \quad \forall\, x, y \in \mathfrak{E}$,

(ii) $\vec{xy} = \vec{xz} + \vec{zy} \quad \forall\, x, y, z \in \mathfrak{E}$, (86)

(iii) given a point o, chosen arbitrarily from $\mathfrak{E}$, there corresponds to each vector $\boldsymbol{x} \in \mathsf{E}$ a unique point $x \in \mathfrak{E}$ such that $\boldsymbol{x} = \vec{ox}$.

In connection with axiom (iii), the point o is called the *origin* and $\boldsymbol{x}$ the *position* of x relative to o.

The notions of distance and angle in $\mathfrak{E}$ are derived from the scalar product on the supporting vector space E: the distance xy between the arbitrary points x and y is defined by $xy = |\vec{xy}|$ and the angle φ subtended by x and y at a third arbitrary point z by[13]

$$\varphi = \cos^{-1}\{(\vec{zx}) \cdot (\vec{zy})/(zx)(zy)\} \quad (0 \leqslant \varphi \leqslant \pi).$$

The properties ascribed to E in Section 1 and to $\mathfrak{E}$ above also allow the analytical geometry of points, lines and planes in $\mathfrak{E}$ to be treated vectorially and the reader is assumed to be acquainted with these developments.

If we choose an origin o in $\mathfrak{E}$ and a basis $g = \{\boldsymbol{g}_1, \boldsymbol{g}_2, \boldsymbol{g}_3\}$ of E, consisting of unit vectors, there is associated with an arbitrary point x in $\mathfrak{E}$ a unique position $\boldsymbol{x}$ in E and a unique ordered triplet of scalars (x_1, x_2, x_3) which are the components of $\boldsymbol{x}$ relative to g. (x_1, x_2, x_3) are called the *coordinates* of x in the coordinate system (o, g) consisting of the origin o and the basis g. When g is an orthonormal basis, (o, g) is said to be a *rectangular Cartesian coordinate system.* All the coordinates employed in the later chapters belong to systems of this kind.

Problem 17[14] Let $\boldsymbol{x}$ be the position of an arbitrary point x relative to an origin o and let $\boldsymbol{S}$ and $\boldsymbol{Q}$ be respectively a positive definite symmetric tensor and a proper orthogonal tensor on E. Give geometrical interpretations of the actions of $\boldsymbol{S}$ and $\boldsymbol{Q}$ on $\boldsymbol{x}$.

Solution. (i) $\boldsymbol{S}$ admits the spectral representation (59), the proper numbers λ_i $(i = 1, 2, 3)$ being positive and the associated proper vectors forming an orthonormal basis $p = \{\boldsymbol{p}_1, \boldsymbol{p}_2, \boldsymbol{p}_3\}$ of E. Hence

$$\boldsymbol{S}\boldsymbol{x} = \sum_{r=1}^{3} \lambda_r(\boldsymbol{p}_r \otimes \boldsymbol{p}_r)\boldsymbol{x} = \sum_{r=1}^{3} \lambda_r(\boldsymbol{x} \cdot \boldsymbol{p}_r)\boldsymbol{p}_r = \sum_{r=1}^{3} \lambda_r x_r \boldsymbol{p}_r,$$

where x_i are the coordinates of x in the rectangular Cartesian system (o, p). In geometrical terms, therefore, the action of $\boldsymbol{S}$ on $\boldsymbol{x}$ is to map x into the point y having position $\boldsymbol{S}\boldsymbol{x}$ relative to o and coordinates $\lambda_i x_i$ in the system (o, p). The distance of every point of $\mathfrak{E}$ from the coordinate plane $x_1 = 0$ is changed by the positive factor λ_1 (being

[13] If $\boldsymbol{a}$ and $\boldsymbol{b}$ are arbitrary vectors, it follows from axioms (7) and (3) that

$$-|\boldsymbol{a}|\,|\boldsymbol{b}| \leqslant \boldsymbol{a} \cdot \boldsymbol{b} \leqslant |\boldsymbol{a}|\,|\boldsymbol{b}|.$$

[14] In an initial reading of Chapter 1 this problem may be omitted.

increased if $\lambda_1 > 1$ and reduced if $0 < \lambda_1 < 1$), and distances from the planes $x_2 = 0$ and $x_3 = 0$ through o are likewise multiplied by λ_2 and λ_3 respectively. The tensor $\boldsymbol{S}$ accordingly gives rise to a transformation of $\mathfrak{E}$ consisting of proportional extensions, or stretches, of amounts λ_i in the mutually orthogonal directions defined by the unit proper vectors $\boldsymbol{p}_i$. These directions are known as the *principal axes* of $\boldsymbol{S}$.

(ii) $\boldsymbol{Q}$ can be expressed in the form (79) where $-\pi < \theta \leqslant \pi$ and $\{\boldsymbol{p}, \boldsymbol{q}, \boldsymbol{r}\}$ is an orthonormal basis of E. Hence

$$\begin{aligned} \boldsymbol{Q}\boldsymbol{x} &= \{\boldsymbol{p} \otimes \boldsymbol{p} + (\boldsymbol{q} \otimes \boldsymbol{q} + \boldsymbol{r} \otimes \boldsymbol{r}) \cos\theta - (\boldsymbol{q} \otimes \boldsymbol{r} - \boldsymbol{r} \otimes \boldsymbol{q}) \sin\theta\}\boldsymbol{x} \\ &= p\boldsymbol{p} + (q \cos\theta - r \sin\theta)\boldsymbol{q} + (q \sin\theta + r \cos\theta)\boldsymbol{r} \end{aligned}$$

with $p = \boldsymbol{x} \cdot \boldsymbol{p}$, $q = \boldsymbol{x} \cdot \boldsymbol{q}$, $r = \boldsymbol{x} \cdot \boldsymbol{r}$. In the rectangular Cartesian

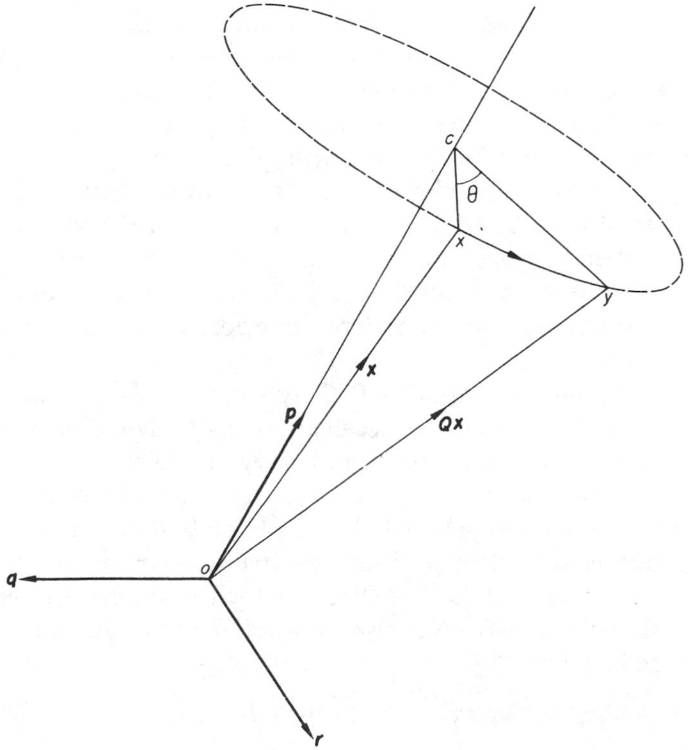

FIGURE 1 Action of a rotation of amount θ about the axis oc on the position of an arbitrary point x (Problem 17).

coordinate system with origin o and base vectors $\boldsymbol{p}, \boldsymbol{q}, \boldsymbol{r}$, x and the point y with position $\boldsymbol{Qx}$ relative to o have coordinates (p, Re z, Im z) and (p, Re $ze^{i\theta}$, Im $ze^{i\theta}$) respectively, where $z = q + ir$, and we see from Figure 1 that x is carried into y by rotating the radius vector ox through an angle θ about the axis oc. The action of $\boldsymbol{Q}$ on $\boldsymbol{x}$ may therefore be interpreted as a rotation of $\mathfrak{E}$ of amount θ about the axis through o in the direction defined by $\boldsymbol{p}$.

10 SCALAR, VECTOR AND TENSOR FIELDS

A *neighbourhood* of a point x is the set of all points whose distances from x are less than some positive real number. In relation to a subset P of $\mathfrak{E}$ the terms *interior point*, *exterior point* and *boundary point* signify in turn a point possessing a neighbourhood which belongs entirely to P, a point which has a neighbourhood containing no point of P, and a point whose every neighbourhood contains at least one point of P and at least one point not belonging to P. A boundary point itself may or may not belong to P. The totalities of interior and boundary points are called the *interior* and the *boundary* of P respectively, the latter being denoted by ∂P. A point set P is said to be *open* if it coincides with its interior, *connected* if every pair of its members can be joined by a path lying wholly in P, and *bounded* if one of its members has a neighbourhood containing every point of P. A *domain* of $\mathfrak{E}$ is a connected open point set and a *region* of $\mathfrak{E}$ is a connected point set with non-empty interior which contains the whole of its boundary.

If to each point of a domain D there corresponds a scalar, the function from D to R so defined is called a *scalar field* on D. Similarly, functions from D to E and from D to L assigning to each point of D a vector and a tensor in turn are called *vector* and *tensor fields* on D.

When an origin o is selected the points of D stand in bijective correspondence with their positions relative to o and we adopt the notation D_0 for the set $\{\boldsymbol{x}: x \in D\}$ of all such positions. Evidently a scalar field $\phi: D \to \mathsf{R}$ gives rise to a scalar-valued function ϕ_0: $D_0 \to \mathsf{R}$ defined by

$$\phi_0(\boldsymbol{x}) = \phi(x) \quad \forall\, \boldsymbol{x} \in D_0,$$

and vector- and tensor-valued functions $\boldsymbol{u}_0: D_0 \to \mathsf{E}$ and $\boldsymbol{T}_0: D_0 \to \mathsf{L}$ correspond in a like manner to vector and tensor fields $\boldsymbol{u}$ and $\boldsymbol{T}$

defined on D. Although the functions ϕ_0, $\boldsymbol{u}_0$ and $\boldsymbol{T}_0$ depend upon the choice of origin which has been made, this is of no consequence since, in view of (86), the effect of a change of origin is merely to increment all positions by a fixed vector. We shall therefore make no distinction henceforth between ϕ and ϕ_0, $\boldsymbol{u}$ and $\boldsymbol{u}_0$ and $\boldsymbol{T}$ and $\boldsymbol{T}_0$, and $\phi(\boldsymbol{x})$, $\boldsymbol{u}(\boldsymbol{x})$ and $\boldsymbol{T}(\boldsymbol{x})$ will denote the values of ϕ, $\boldsymbol{u}$ and $\boldsymbol{T}$ at the representative point x with position $\boldsymbol{x}$ relative to an arbitrarily chosen origin o. If an orthonormal basis e is adjoined to o, constituting a rectangular Cartesian coordinate system in which the typical point x has coordinates (x_1, x_2, x_3), ϕ, $\boldsymbol{u}$ and $\boldsymbol{T}$ and the components of $\boldsymbol{u}$ and $\boldsymbol{T}$ relative to e may be regarded as functions of the three variables x_1, x_2, x_3.

As well as being functions of position, scalar, vector and tensor fields may depend upon one or more real parameters. This is commonly the case in continuum mechanics where field quantities describing the behaviour of a moving body are apt to vary with time. The results given in the remainder of this chapter are unaffected by the dependence upon parameters of the fields concerned.

A scalar field ϕ, defined on a domain D, is said to be *continuous* if

$$\lim_{\alpha \to 0} |\phi(\boldsymbol{x} + \alpha \boldsymbol{a}) - \phi(\boldsymbol{x})| = 0 \quad \forall\, \boldsymbol{x} \in D_0, \boldsymbol{a} \in \mathsf{E},$$

and *differentiable* if there exists a vector field $\boldsymbol{w}$ such that

$$\lim_{\alpha \to 0} |\boldsymbol{w}(\boldsymbol{x}) \,.\, \boldsymbol{a} - \alpha^{-1}\{\phi(\boldsymbol{x} + \alpha \boldsymbol{a}) - \phi(\boldsymbol{x})\}| = 0 \quad \forall\, \boldsymbol{x} \in D_0, \boldsymbol{a} \in \mathsf{E}. \tag{87}$$

There is at most one vector field with the property (87); for if there were two, $\boldsymbol{w}$ and $\overline{\boldsymbol{w}}$ say, it would follow from the inequalities

$$0 \leqslant |\{\boldsymbol{w}(\boldsymbol{x}) - \overline{\boldsymbol{w}}(\boldsymbol{x})\} \,.\, \boldsymbol{a}| \leqslant |\boldsymbol{w}(\boldsymbol{x}) \,.\, \boldsymbol{a} - \alpha^{-1}\{\phi(\boldsymbol{x} + \alpha \boldsymbol{a}) - \phi(\boldsymbol{x})\}|$$
$$+ |\overline{\boldsymbol{w}}(\boldsymbol{x}) \,.\, \boldsymbol{a} - \alpha^{-1}\{\phi(\boldsymbol{x} + \alpha \boldsymbol{a}) - \phi(\boldsymbol{x})\}|,$$

on letting $\alpha \to 0$ and exploiting the arbitrariness of $\boldsymbol{a}$, that $\overline{\boldsymbol{w}} = \boldsymbol{w}$. When ϕ is differentiable the unique vector field $\boldsymbol{w}$ is called the *gradient* of ϕ and denoted by grad ϕ.

The properties of continuity and differentiability are attributed to a vector field $\boldsymbol{u}$ and a tensor field $\boldsymbol{T}$ defined on D if they apply to the scalar fields $\boldsymbol{u} \,.\, \boldsymbol{a}$ and $\boldsymbol{a} \,.\, (\boldsymbol{T}\boldsymbol{b})$ for all choices of the vectors $\boldsymbol{a}$ and $\boldsymbol{b}$.[15] If grad ϕ exists and is continuous, ϕ is said to be *continuously*

[15] Note that in this statement $\boldsymbol{a}$ and $\boldsymbol{b}$ are arbitrary vectors in E, *not* vector fields on D.

differentiable, and this property extends to $\boldsymbol{u}$ and $\boldsymbol{T}$ if grad$(\boldsymbol{u} \,.\, \boldsymbol{a})$ and grad$\{\boldsymbol{a} \,.\, (\boldsymbol{T}\boldsymbol{b})\}$ exist and are continuous for all $\boldsymbol{a}, \boldsymbol{b} \in \mathsf{E}$.

Given that the vector field $\boldsymbol{u}$ is differentiable, the gradient of $\boldsymbol{u}$, denoted by grad $\boldsymbol{u}$, is the tensor field[16] defined by

$$\{(\text{grad}\, \boldsymbol{u})(\boldsymbol{x})\}^{\mathrm{T}}\boldsymbol{a} = \text{grad}\, \{\boldsymbol{u}(\boldsymbol{x}) \,.\, \boldsymbol{a}\} \quad \forall \boldsymbol{x} \in D_0, \boldsymbol{a} \in \mathsf{E}. \tag{88}$$

The *divergence* and the *curl* of $\boldsymbol{u}$, written div $\boldsymbol{u}$ and curl $\boldsymbol{u}$, are respectively scalar- and vector-valued and are defined by

$$(\text{div}\, \boldsymbol{u})(\boldsymbol{x}) = \text{tr}\{(\text{grad}\, \boldsymbol{u})(\boldsymbol{x})\} \quad \forall\, \boldsymbol{x} \in D_0, \tag{89}$$

$$\{(\text{curl}\, \boldsymbol{u})(\boldsymbol{x})\} \,.\, \boldsymbol{a} = \text{div}\, \{\boldsymbol{u}(\boldsymbol{x}) \wedge \boldsymbol{a}\} \quad \forall\, \boldsymbol{x} \in D_0, \boldsymbol{a} \in \mathsf{E}. \tag{90}$$

When the tensor field $\boldsymbol{T}$ is differentiable, its divergence, denoted by div $\boldsymbol{T}$, is the vector field defined by

$$\{(\text{div}\, \boldsymbol{T})(\boldsymbol{x})\} \,.\, \boldsymbol{a} = \text{div}\{\boldsymbol{T}(\boldsymbol{x})\, \boldsymbol{a}\} \quad \forall\, \boldsymbol{x} \in D_0, \boldsymbol{a} \in \mathsf{E}. \tag{91}$$

Problem 18 Let ϕ, $\boldsymbol{u}$ and $\boldsymbol{T}$ be differentiable scalar, vector and tensor fields. Show that, with reference to the rectangular Cartesian coordinate system (o, e), where $e = \{\boldsymbol{e}_1, \boldsymbol{e}_2, \boldsymbol{e}_3\}$,

$$\text{grad}\, \phi = \frac{\partial \phi}{\partial x_p} \boldsymbol{e}_p, \tag{92}$$

$$\text{grad}\, \boldsymbol{u} = \frac{\partial u_p}{\partial x_q} \boldsymbol{e}_p \otimes \boldsymbol{e}_q, \quad \text{div}\, \boldsymbol{u} = \frac{\partial u_p}{\partial x_p}, \quad \text{curl}\, \boldsymbol{u} = \pm \varepsilon_{pqr} \frac{\partial u_r}{\partial x_q} \boldsymbol{e}_p, \tag{93}$$

$$\text{div}\, \boldsymbol{T} = \frac{\partial T_{pq}}{\partial x_p} \boldsymbol{e}_q. \tag{94}$$

Here u_i and T_{ij} are the components of $\boldsymbol{u}$ and $\boldsymbol{T}$ relative to e and the upper or lower sign holds in $(93)_3$ according as e is positive in E^+ or E^-.

Solution. Since ϕ is differentiable it follows from equation (87), on replacing $\boldsymbol{a}$ by the base vectors $\boldsymbol{e}_i$ in turn, that the partial derivatives

[16] Concerning the existence of a unique tensor field satisfying the definition (88) and unique vector fields satisfying (90) and (91), see the remarks in parentheses at the end of Problem 18 below.

$\partial\phi/\partial x_i$ exist in D and that, moreover, $w_i = \partial\phi/\partial x_i$. Hence

$$\text{grad}\,\phi = \boldsymbol{w} = w_p \boldsymbol{e}_p = \frac{\partial\phi}{\partial x_p}\boldsymbol{e}_p.$$

The differentiability of $\boldsymbol{u}$ and $\boldsymbol{T}$ ensures the existence in D of the partial derivatives $\partial u_i/\partial x_j$ and $\partial T_{ij}/\partial x_k$. Equations (93) and (94) are established, by appeal to the definitions (88) to (91) and with use of (92), by the following calculations.

$$\left(\text{grad}\,\boldsymbol{u} - \frac{\partial u_p}{\partial x_q}\boldsymbol{e}_p \otimes \boldsymbol{e}_q\right)^{\mathrm{T}} \boldsymbol{a} = (\text{grad}\,\boldsymbol{u})^{\mathrm{T}}\boldsymbol{a} - \frac{\partial u_p}{\partial x_q}(\boldsymbol{e}_q \otimes \boldsymbol{e}_p)\boldsymbol{a}$$

$$= \text{grad}\,(\boldsymbol{u}\,.\,\boldsymbol{a}) - \frac{\partial u_p}{\partial x_q} a_p \boldsymbol{e}_q = \text{grad}\,(\boldsymbol{u}\,.\,\boldsymbol{a}) - \frac{\partial}{\partial x_q}(\boldsymbol{u}\,.\,\boldsymbol{a})\boldsymbol{e}_q = \boldsymbol{0}. \qquad \text{(A)}$$

$$\text{div}\,\boldsymbol{u} = \text{tr}\left(\frac{\partial u_p}{\partial x_q}\boldsymbol{e}_p \otimes \boldsymbol{e}_q\right) = \frac{\partial u_p}{\partial x_q}\text{tr}\,(\boldsymbol{e}_p \otimes \boldsymbol{e}_q) = \frac{\partial u_p}{\partial x_q}\boldsymbol{e}_p\,.\,\boldsymbol{e}_q$$

$$= \frac{\partial u_p}{\partial x_q}\delta_{pq} = \frac{\partial u_p}{\partial x_p}.$$

$$\left(\text{curl}\,\boldsymbol{u} \mp \varepsilon_{pqr}\frac{\partial u_r}{\partial x_q}\boldsymbol{e}_p\right).\,\boldsymbol{a} = (\text{curl}\,\boldsymbol{u})\,.\,\boldsymbol{a} \mp \varepsilon_{qrp}\frac{\partial u_r}{\partial x_q}a_p$$

$$= \text{div}(\boldsymbol{u} \wedge \boldsymbol{a}) - \frac{\partial}{\partial x_q}(\pm\varepsilon_{qrp}u_r a_p) = 0. \qquad \text{(B)}$$

$$\left(\text{div}\,\boldsymbol{T} - \frac{\partial T_{pq}}{\partial x_p}\boldsymbol{e}_q\right).\,\boldsymbol{a} = (\text{div}\,\boldsymbol{T})\,.\,\boldsymbol{a} - \frac{\partial T_{pq}}{\partial x_p}a_q$$

$$= \text{div}(\boldsymbol{T}\boldsymbol{a}) - \frac{\partial}{\partial x_p}(T_{pq}a_q) = 0. \qquad \text{(C)}$$

[A rearrangement of the steps in (A) confirms that the tensor on the right of $(93)_1$ satisfies the definition (88), and it is an easy matter to show that no other tensor field meets this requirement. Similarly, the existence of vector fields satisfying the definitions (90) and (91) can be proved by recasting (B) and (C), and again it is a simple task to verify that these fields are unique.]

The reader is assumed to have some familiarity with the elementary differential geometry of curves and surfaces and with the basic theory of line, surface and volume integrals. The notation $\mathrm{d}\boldsymbol{x}$ is used for the directed element of arc length on a curve in $\mathfrak{E}$, and the differentials of area on a surface in $\mathfrak{E}$ and of volume in a region of $\mathfrak{E}$ are denoted by $\mathrm{d}a$ and $\mathrm{d}v$ respectively.

Problem 19 Let ϕ be a scalar field which is continuous in a domain D. If

$$\int_R \phi \,\mathrm{d}v = 0$$

for every region R contained in D, prove that $\phi = 0$ in D.

Solution. Suppose that the stated conclusion is false. Then there is a point y in D, with position $\boldsymbol{y}$, say, relative to some origin o, at which ϕ is non-zero. Let $\phi(\boldsymbol{y}) = 2\kappa$ and assume that $\kappa > 0$. Since y is necessarily an interior point (D being open), every point within some positive distance δ of y is in D. And because ϕ is continuous at y it is possible to choose a positive real number $\delta' < \delta$ such that

$$|\phi(\boldsymbol{x}) - \phi(\boldsymbol{y})| < \kappa \quad \text{whenever} \quad |\boldsymbol{x} - \boldsymbol{y}| < \delta'.$$

Now the point set $B = \{x: xy \leqslant \delta'\}$ is a region contained in D and at the representative point $x \in B$, with position $\boldsymbol{x}$ relative to o, $\phi(\boldsymbol{x}) > \phi(\boldsymbol{y}) - \kappa = \kappa$. Hence

$$\int_B \phi \,\mathrm{d}v > \kappa \int_B \mathrm{d}v = \tfrac{4}{3}\pi\delta'^3\kappa > 0,$$

which is contrary to hypothesis. The assumption that $\kappa < 0$ similarly leads to a contradiction, so there is no point in D at which ϕ is non-zero.

[The result which has just been proved continues to hold when ϕ is replaced by a vector or tensor field. It can also be shown, by a straightforward modification of the reasoning used above, that (i) a continuous scalar field ϕ such that

$$\int_S \phi \,\mathrm{d}a = 0$$

for every surface segment S contained in D vanishes in D, and (ii) either of the conditions

$$\int_S \boldsymbol{u}\,\mathrm{d}a = \mathbf{0} \quad \text{or} \quad \int_S \boldsymbol{u}\,.\,\boldsymbol{n}\,\mathrm{d}a = 0,$$

holding for every surface segment S in D, suffices for the vanishing in D of a continuous vector field $\boldsymbol{u}$, $\boldsymbol{n}$ being a unit vector normal to S.]

11 INTEGRAL THEOREMS

The transformation of an integral over the boundary ∂R of a region R into an integral over R is an essential step in the derivation of the field equations of continuum mechanics. The means by which it is accomplished is the *divergence theorem* and a form of this result sufficiently general to meet all later requirements can be stated as follows.

Let R be a regular region of $\mathfrak{E}$ with boundary ∂R, let $\boldsymbol{n}$ be the outward unit vector normal to ∂R, and let $\boldsymbol{u}$ be a vector field and $\boldsymbol{T}$ a tensor field, each continuous in R and continuously differentiable in the interior of R. Then

$$\int_{\partial R} \boldsymbol{u} \otimes (\boldsymbol{T}^{\mathrm{T}}\boldsymbol{n})\,\mathrm{d}a = \int_R \{\boldsymbol{u} \otimes \operatorname{div} \boldsymbol{T} + (\operatorname{grad} \boldsymbol{u})\boldsymbol{T}\}\,\mathrm{d}v. \tag{95}$$

Implied in the term *regular region* are certain restrictions of a geometrical kind on the boundary ∂R which are not easy to characterize concisely.[17] Broadly speaking, however, ∂R must be a bounded, two-sided (i.e. orientable[18]) surface divisible into a finite number of *regular surface segments*, this latter term alluding to an orientable surface possessing a continuous field of unit normal vectors.

We take note of two special cases of the above formula. First, regarding $\boldsymbol{u}$ as a fixed vector, equation (95) becomes

$$\boldsymbol{u} \otimes \int_{\partial R} \boldsymbol{T}^{\mathrm{T}}\boldsymbol{n}\,\mathrm{d}a = \boldsymbol{u} \otimes \int_R \operatorname{div} \boldsymbol{T}\,\mathrm{d}v,$$

[17] See O. D. Kellogg, *Foundations of Potential Theory* (New York, Ungar, 1929), Chapter IV, or, for a more modern treatment, H. Flanders, *Differential Forms with Applications to the Physical Sciences* (New York etc., Academic Press, 1963), Chapter V.

[18] See, for example, T. M. Apostol, *Mathematical Analysis* (Reading, Mass. etc., Addison-Wesley, 1963), p. 338.

and since the choice of $\boldsymbol{u}$ from E is unrestricted it follows that

$$\int_{\partial R} \boldsymbol{T}^{\mathrm{T}}\boldsymbol{n}\, \mathrm{d}a = \int_R \operatorname{div} \boldsymbol{T}\, \mathrm{d}v. \tag{96}$$

Secondly, if $\boldsymbol{T} = \boldsymbol{I}$, (95) reduces to

$$\int_{\partial R} \boldsymbol{u} \otimes \boldsymbol{n}\, \mathrm{d}a = \int_R \operatorname{grad} \boldsymbol{u}\, \mathrm{d}v. \tag{97}$$

On forming the trace of each side of (97) and making use of equations $(47)_1$ and (89) we obtain

$$\int_{\partial R} \boldsymbol{u} \cdot \boldsymbol{n}\, \mathrm{d}a = \int_R \operatorname{div} \boldsymbol{u}\, \mathrm{d}v, \tag{98}$$

which is the standard form of the divergence theorem. It might thus appear that (95) is a more general result than (98), but in fact (95), and hence (96) and (97), can be deduced from (98). To see this let the vector field $(\boldsymbol{u} \cdot \boldsymbol{a})(\boldsymbol{T}\boldsymbol{b})$ replace $\boldsymbol{u}$ in (98), $\boldsymbol{a}$ and $\boldsymbol{b}$ being arbitrary vectors. The identities

$$\{(\boldsymbol{u} \cdot \boldsymbol{a})(\boldsymbol{T}\boldsymbol{b})\} \cdot \boldsymbol{n} = \boldsymbol{a} \cdot [\{\boldsymbol{u} \otimes (\boldsymbol{T}^{\mathrm{T}}\boldsymbol{n})\}\boldsymbol{b}],$$

$$\operatorname{div}\{(\boldsymbol{u} \cdot \boldsymbol{a})(\boldsymbol{T}\boldsymbol{b})\} = \boldsymbol{a} \cdot [\{(\operatorname{grad} \boldsymbol{u})\boldsymbol{T} + \boldsymbol{u} \otimes \operatorname{div} \boldsymbol{T}\}\boldsymbol{b}],$$

are readily derived with the aid of the definitions (24), (41), (88) and (91) and the well-known formula

$$\operatorname{div}(\phi \boldsymbol{u}) = \boldsymbol{u} \cdot \operatorname{grad} \phi + \phi \operatorname{div} \boldsymbol{u}.$$

Employing them in the amended form of (98) and appealing to the arbitrariness of $\boldsymbol{a}$ and $\boldsymbol{b}$ we recover equation (95).

Problem 20 Let R be a regular region and $\boldsymbol{T}$ a tensor field which is continuous in R and continuously differentiable in the interior of R. Show that

$$\int_{\partial R} \boldsymbol{x} \wedge (\boldsymbol{T}^{\mathrm{T}}\boldsymbol{n})\, \mathrm{d}a = \int_R (\boldsymbol{x} \wedge \operatorname{div} \boldsymbol{T} - \boldsymbol{\tau})\, \mathrm{d}v, \tag{99}$$

where $\boldsymbol{x}$ is the position of a representative point of R and $\boldsymbol{\tau}$ is the axial vector of $\boldsymbol{T} - \boldsymbol{T}^{\mathrm{T}}$.

Solution. On replacing $\boldsymbol{b}$ by $\boldsymbol{x}$ and $\boldsymbol{c}$ by $\boldsymbol{T}^{\mathrm{T}}\boldsymbol{n}$ in the vector identity

$(73)_1$ and then integrating each side over the surface ∂R we obtain

$$\boldsymbol{a} \wedge \int_{\partial R} \boldsymbol{x} \wedge (\boldsymbol{T}^{\mathrm{T}}\boldsymbol{n})\,\mathrm{d}a = \left(\int_{\partial R} \{\boldsymbol{x} \otimes (\boldsymbol{T}^{\mathrm{T}}\boldsymbol{n}) - (\boldsymbol{T}^{\mathrm{T}}\boldsymbol{n}) \otimes \boldsymbol{x}\}\,\mathrm{d}a\right)\boldsymbol{a}, \tag{A}$$

$\boldsymbol{a}$ being an arbitrary vector. Next, setting $\boldsymbol{u} = \boldsymbol{x}$ in the divergence theorem (95) and noting that $\operatorname{grad} \boldsymbol{x} = \boldsymbol{I}$, we have

$$\int_{\partial R} \boldsymbol{x} \otimes (\boldsymbol{T}^{\mathrm{T}}\boldsymbol{n})\,\mathrm{d}a = \int_{R} (\boldsymbol{x} \otimes \operatorname{div} \boldsymbol{T} + \boldsymbol{T})\,\mathrm{d}v. \tag{B}$$

When transformed into a volume integral with the aid of (B) the right-hand side of (A) becomes

$$\int_{R} \{(\boldsymbol{x} \otimes \operatorname{div} \boldsymbol{T} - \operatorname{div} \boldsymbol{T} \otimes \boldsymbol{x})\boldsymbol{a} + (\boldsymbol{T} - \boldsymbol{T}^{\mathrm{T}})\boldsymbol{a}\}\,\mathrm{d}v. \tag{C}$$

A second application of the identity $(73)_1$ puts the integrand of (C) into the form

$$\boldsymbol{a} \wedge (\boldsymbol{x} \wedge \operatorname{div} \boldsymbol{T}) + \boldsymbol{\tau} \wedge \boldsymbol{a},$$

use also being made of equation (71) as applied to the skew-symmetric tensor $\boldsymbol{T} - \boldsymbol{T}^{\mathrm{T}}$. Hence

$$\boldsymbol{a} \wedge \int_{\partial R} \boldsymbol{x} \wedge (\boldsymbol{T}^{\mathrm{T}}\boldsymbol{n})\,\mathrm{d}a = \boldsymbol{a} \wedge \int_{R} (\boldsymbol{x} \wedge \operatorname{div} \boldsymbol{T} - \boldsymbol{\tau})\,\mathrm{d}v,$$

and equation (99) follows on account of the arbitrariness of $\boldsymbol{a}$.

Also of basic importance in continuum mechanics is the conversion of a line integral around a closed curve into an integral over an open surface spanning the curve. The requisite transformation is provided by *Stokes's theorem*, the following form[19] of which is suited to subsequent applications.

Let Λ be a regular surface segment having as its boundary a simple closed curve Γ and let $\boldsymbol{u}$ be a vector field which is continuously differentiable in some domain containing Λ. Then

$$\oint_{\Gamma} \boldsymbol{u} \,.\, \mathrm{d}\boldsymbol{x} = \int_{\Lambda} \operatorname{curl} \boldsymbol{u} \,.\, \boldsymbol{n}\,\mathrm{d}a, \tag{100}$$

[19] For a proof, see T. M. Apostol, *op. cit.* p. 335.

where Γ has positive orientation relative to the unit vector field $\boldsymbol{n}$ *normal to Λ.*

The reference in the foregoing statement to the orientation (or sense of description) of Γ is illustrated in Figure 2. x, y and z are

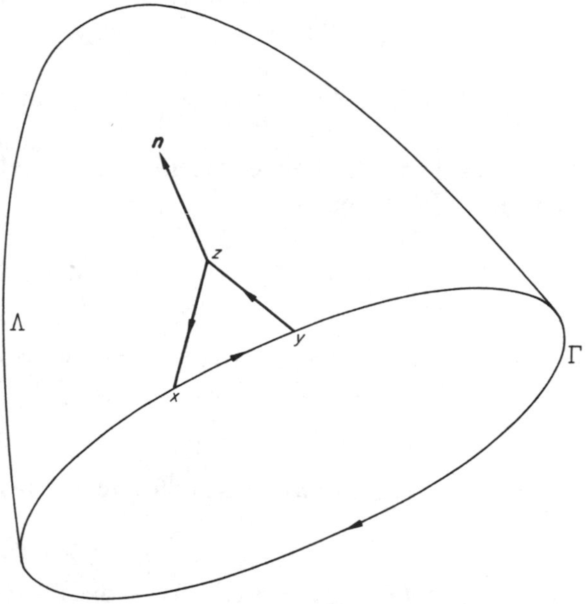

FIGURE 2 Orientation of a simple closed curve Γ relative to a unit normal vector field $\boldsymbol{n}$ on a spanning surface Λ.

nearby points, x and y on Γ and z an interior point of Λ. The indicated sense of description of the circuit xyz, induced by the orientation of Γ, is related to the direction of the unit normal to Λ at z by a right-hand screw rule.[20]

EXERCISES

(The starred exercises contain results which are used in the later chapters.)

[20] In more formal terms, there exists a positive real number ε such that $[\vec{xz}, \vec{yz}, \boldsymbol{n}] > 0$ whenever xz and yz are less than ε.

1*. Show that an arbitrary tensor $\boldsymbol{A}$ can be expressed as the sum of a *spherical tensor* (i.e. a scalar multiple of the identity tensor) and a tensor with zero trace. Prove that this decomposition is unique and that $\boldsymbol{A}'$, the traceless part of $\boldsymbol{A}$, is given by

$$\boldsymbol{A}' = \boldsymbol{A} - \tfrac{1}{3}(\operatorname{tr}\boldsymbol{A})\boldsymbol{I}.$$

[$\boldsymbol{A}'$ is called the *deviator* of $\boldsymbol{A}$.]

2*. Let $\boldsymbol{A}$ be an arbitrary tensor. Show that

$$II_{\boldsymbol{A}} = \tfrac{1}{2}\{(\operatorname{tr}\boldsymbol{A})^2 - \operatorname{tr}\boldsymbol{A}^2\}.$$

Using the Cayley–Hamilton theorem, deduce that

$$III_{\boldsymbol{A}} = \tfrac{1}{6}\{(\operatorname{tr}\boldsymbol{A})^3 - 3\operatorname{tr}\boldsymbol{A}\operatorname{tr}\boldsymbol{A}^2 + 2\operatorname{tr}\boldsymbol{A}^3\}.$$

3. Using the result

$$\det\boldsymbol{A}\det\boldsymbol{B} = \det(\boldsymbol{A}^{\mathrm{T}}\boldsymbol{B}) \quad \forall\, \boldsymbol{A}, \boldsymbol{B} \in \mathsf{L},$$

or otherwise, show that

$$\varepsilon_{ijk}\varepsilon_{lmn} = \delta_{il}(\delta_{jm}\delta_{kn} - \delta_{jn}\delta_{km}) + \delta_{im}(\delta_{jn}\delta_{kl} - \delta_{jl}\delta_{kn}) + \delta_{in}(\delta_{jl}\delta_{km} - \delta_{jm}\delta_{kl}).$$

Hence derive the formulae

(*a*) $\varepsilon_{ijp}\varepsilon_{lmp} = \delta_{il}\delta_{jm} - \delta_{im}\delta_{jl}$,

(*b*) $\varepsilon_{ipq}\varepsilon_{lpq} = 2\delta_{il}$.

4. Let $\boldsymbol{A}$ be an arbitrary tensor and $\boldsymbol{A}^*$ its adjugate.

(i) Given that A_{ij} are the components of $\boldsymbol{A}$ relative to an orthonormal basis *e*, show that the components of $\boldsymbol{A}^*$ are $\frac{1}{2}\varepsilon_{ipq}\varepsilon_{jrs}A_{pr}A_{qs}$. Deduce that $\boldsymbol{A}^{\mathrm{T}*} = \boldsymbol{A}^{*\mathrm{T}}$.

(ii) Show that

(*a*) $(\boldsymbol{A}^*)^* = (\det\boldsymbol{A})\boldsymbol{A}$,

(*b*) $\operatorname{tr}\boldsymbol{A}^* = II_{\boldsymbol{A}}$,

(*c*) $\boldsymbol{A}\{\boldsymbol{a} \wedge (\boldsymbol{A}^{\mathrm{T}}\boldsymbol{b})\} = (\boldsymbol{A}^*\boldsymbol{a}) \wedge \boldsymbol{b} \quad \forall\, \boldsymbol{a}, \boldsymbol{b} \in \mathsf{E}$.

5. Let $\boldsymbol{A}$ and $\boldsymbol{B}$ be arbitrary tensors, $\boldsymbol{A}^*$ and $\boldsymbol{B}^*$ the adjugates of $\boldsymbol{A}$

and $\boldsymbol{B}$, and α and β arbitrary scalars. Show that

$$\det(\alpha\boldsymbol{A} + \beta\boldsymbol{B}) = \alpha^3 \det \boldsymbol{A} + \alpha^2\beta \operatorname{tr}(\boldsymbol{B}^{\mathrm{T}}\boldsymbol{A}^*) + \alpha\beta^2 \operatorname{tr}(\boldsymbol{A}^{\mathrm{T}}\boldsymbol{B}^*) + \beta^3 \det \boldsymbol{B}.$$

6. Relative to an orthonormal basis e the components of a symmetric tensor $\boldsymbol{S}$ are given by

$$-S_{11} = S_{33} = \tfrac{1}{3}, \quad -S_{12} = -S_{21} = S_{23} = S_{32} = \tfrac{2}{3},$$
$$S_{22} = S_{31} = S_{13} = 0.$$

Find the proper numbers of $\boldsymbol{S}$ and the components relative to e of an orthonormal set of associated proper vectors.

7. Let $\boldsymbol{p}$ be a non-zero vector and $\boldsymbol{S}$ a symmetric tensor. Prove that $\boldsymbol{p}$ is a proper vector of $\boldsymbol{S}$ if and only if $\boldsymbol{p} \otimes \boldsymbol{p}$ commutes with $\boldsymbol{S}$.

8. (i) Let $\boldsymbol{W}$ be a skew-symmetric tensor with axial vector $\boldsymbol{w}$. Express the components of $\boldsymbol{W}$ (relative to an orthonormal basis) in terms of the components of $\boldsymbol{w}$ and obtain the inverse relations.

(ii)* If $\boldsymbol{S}$ is a symmetric and $\boldsymbol{W}$ a skew-symmetric tensor, prove that $\operatorname{tr}(\boldsymbol{SW}) = 0$.

9. Let $\boldsymbol{W}_1$ and $\boldsymbol{W}_2$ be skew-symmetric tensors with axial vectors $\boldsymbol{w}_1$ and $\boldsymbol{w}_2$ respectively. Show that

(a) $\boldsymbol{W}_1\boldsymbol{W}_2 = \boldsymbol{w}_2 \otimes \boldsymbol{w}_1 - (\boldsymbol{w}_1 \cdot \boldsymbol{w}_2)\boldsymbol{I}$,

(b) $\operatorname{tr}(\boldsymbol{W}_1\boldsymbol{W}_2) = -2\boldsymbol{w}_1 \cdot \boldsymbol{w}_2$.

By making appropriate choices of $\boldsymbol{W}_1$ and $\boldsymbol{W}_2$, deduce from (b) the vector identity

$$(\boldsymbol{a} \wedge \boldsymbol{b}) \cdot (\boldsymbol{c} \wedge \boldsymbol{d}) = (\boldsymbol{a} \cdot \boldsymbol{c})(\boldsymbol{b} \cdot \boldsymbol{d}) - (\boldsymbol{a} \cdot \boldsymbol{d})(\boldsymbol{b} \cdot \boldsymbol{c}) \quad \forall\, \boldsymbol{a}, \boldsymbol{b}, \boldsymbol{c}, \boldsymbol{d} \in \mathsf{E}.$$

10*. Let $e = \{\boldsymbol{e}_1, \boldsymbol{e}_2, \boldsymbol{e}_3\}$ and $e' = \{\boldsymbol{e}'_1, \boldsymbol{e}'_2, \boldsymbol{e}'_3\}$ be orthonormal bases. Show that $\boldsymbol{e}'_p \otimes \boldsymbol{e}_p$ is an orthogonal tensor, proper or improper according as e and e' are similar or opposite. Show also that every orthogonal tensor can be represented in terms of two orthonormal bases in this way.

Let $\boldsymbol{a}$ be an arbitrary vector and $\boldsymbol{A}$ an arbitrary tensor and let their components relative to e and e' be a_i, A_{ij} and a'_i, A'_{ij} respectively. Obtain the transformation rules

$$a'_i = l_{ip}a_p \quad \text{and} \quad A'_{ij} = l_{ip}l_{jq}A_{pq},$$

where $l_{ij} = \boldsymbol{e}'_i \cdot \boldsymbol{e}_j$.

11*. Prove that a tensor which commutes with every proper orthogonal tensor is spherical.

12. Relative to an orthonormal basis *e* the components of a tensor $\boldsymbol{A}$ are given by

$$A_{11} = A_{12} = 2, \quad -A_{21} = A_{22} = -A_{33} = 1,$$
$$A_{23} = A_{32} = A_{31} = A_{13} = 0.$$

Verify that $\boldsymbol{A}$ is invertible and find its polar decompositions relative to *e*.

13. Show that if $\boldsymbol{p}$ is a unit vector, $\boldsymbol{I} - 2\boldsymbol{p} \otimes \boldsymbol{p}$ is an improper orthogonal tensor. Interpret geometrically the action of this tensor on the positions of points of $\mathfrak{E}$.

14*. Let ϕ, $\boldsymbol{u}$ and $\boldsymbol{T}$ be differentiable scalar, vector and tensor fields. Show that

(i) $\operatorname{grad}(\phi \boldsymbol{u}) = \boldsymbol{u} \otimes \operatorname{grad} \phi + \phi \operatorname{grad} \boldsymbol{u}$.

(ii) $\operatorname{div}(\boldsymbol{T}\boldsymbol{u}) = \boldsymbol{u} \cdot \operatorname{div} \boldsymbol{T} + \operatorname{tr}(\boldsymbol{T} \operatorname{grad} \boldsymbol{u})$.

(iii) $\operatorname{div}(\phi \boldsymbol{T}) = \boldsymbol{T}^{\mathrm{T}} \operatorname{grad} \phi + \phi \operatorname{div} \boldsymbol{T}$.

15*. Let $\boldsymbol{u}$ be a vector field which is continuously differentiable in a domain D. Show that $\operatorname{grad} \boldsymbol{u} - (\operatorname{grad} \boldsymbol{u})^{\mathrm{T}}$ is a skew-symmetric tensor field on D with axial vector $\operatorname{curl} \boldsymbol{u}$. Hence, or otherwise, show that

$$\int_{\partial R} \boldsymbol{n} \wedge \boldsymbol{u} \, da = \int_R \operatorname{curl} \boldsymbol{u} \, dv,$$

where R is a regular region contained in D and $\boldsymbol{n}$ the outward unit vector normal to its boundary ∂R.

Chapter 2

BASIC KINEMATICS

Kinematics is the study of motion *per se*, regardless of the forces causing it. The primitive concepts concerned are *position*, *time* and *body*, the latter abstracting into mathematical terms intuitive ideas about aggregations of matter capable of motion and deformation. In the first three sections of this chapter we develop the simple kinematic notions involved in the formulation of the field equations of continuum mechanics. The more elaborate kinematic considerations which are basic to the theoretical mechanics of elastic materials and viscous fluids are then introduced in Sections 4 to 6.

1 BODIES, CONFIGURATIONS AND MOTIONS

A *body* $\mathscr{B}$ is a set whose elements can be put into bijective correspondence with the points of a region B of a Euclidean point space $\mathfrak{E}$. The elements of $\mathscr{B}$ are called *particles*[1] and B is referred to as a *configuration* of $\mathscr{B}$; the point in B to which a given particle of $\mathscr{B}$ corresponds is said to be *occupied* by that particle. If $\mathscr{X}$ denotes a representative particle of $\mathscr{B}$ and $\boldsymbol{x}$ the position relative to an origin o of the point x occupied by $\mathscr{X}$ in B, the preceding statement implies the existence of a function[2] $\boldsymbol{\theta}$: $\mathscr{B} \to B_0$ and its inverse Θ: $B_0 \to \mathscr{B}$ such that

$$\boldsymbol{x} = \boldsymbol{\theta}(\mathscr{X}), \quad \mathscr{X} = \Theta(\boldsymbol{x}). \tag{1}$$

In a motion of a body the configuration changes with time. Let t be a real variable, denoting time, and I an interval (not necessarily bounded) of the reals R. If, with each value of t in I, there is associated a unique configuration B_t of a body $\mathscr{B}$, the family of configurations

[1] As used in continuum mechanics this term is merely a description of an element of a body. It does *not* signify a point mass as in Newtonian mechanics.

[2] B_0 stands for the totality of the positions relative to o of the points of B (see p. 38).

$\{B_t : t \in I\}$ is called a *motion* of $\mathscr{B}$. This definition entails the existence of functions $\boldsymbol{\phi}$: $\mathscr{B} \times I \to (B_t)_0$ and $\boldsymbol{\Phi}$: $\{(\boldsymbol{x}, t)\colon t \in I,\ \boldsymbol{x} \in (B_t)_0\} \to \mathscr{B}$ such that

$$\boldsymbol{x} = \boldsymbol{\phi}(\mathscr{X}, t), \quad \mathscr{X} = \boldsymbol{\Phi}(\boldsymbol{x}, t). \tag{2}$$

In a motion of $\mathscr{B}$ a typical particle $\mathscr{X}$ occupies a succession of points which together form a curve in $\mathfrak{E}$. This curve is called the *path* of $\mathscr{X}$ and is given parametrically by equation $(2)_1$. The *velocity* and the *acceleration* of $\mathscr{X}$ are defined as the rates of change with time of position and velocity respectively as $\mathscr{X}$ traverses its path.

Equations (2) depict a motion as a sequence of correspondences between particles and points. The points are identified by their positions relative to a selected origin o and it is desirable to have a similar means of recognizing particles. To this end a particular configuration B_r of $\mathscr{B}$, distinguished by the name *reference configuration*, is chosen and a typical particle $\mathscr{X}$ labelled by the position $\boldsymbol{X}$ of the point which it occupies in B_r. The origin O from which $\boldsymbol{X}$ is measured need not coincide with o. Equations (1) imply the existence of functions $\boldsymbol{\pi} : \mathscr{B} \to (B_r)_0$ and $\Pi : (B_r)_0 \to \mathscr{B}$ such that $\boldsymbol{X} = \boldsymbol{\pi}(\mathscr{X})$, $\mathscr{X} = \Pi(\boldsymbol{X})$, and on combining these relations with (2) we obtain

$$\boldsymbol{x} = \boldsymbol{\phi}(\Pi(\boldsymbol{X}), t) = \boldsymbol{\psi}(\boldsymbol{X}, t), \tag{3}$$

$$\boldsymbol{X} = \boldsymbol{\pi}(\boldsymbol{\Phi}(\boldsymbol{x}, t)) = \boldsymbol{\Psi}(\boldsymbol{x}, t). \tag{4}$$

Equations (3) and (4), linking the referential and current positions of a typical particle and the time, provide alternative representations of a motion of $\mathscr{B}$. The functions $\boldsymbol{\psi}$ and $\boldsymbol{\Psi}$ are vector fields on B_r and B_t respectively and they depend of course upon the choice of reference configuration.

In order to exclude from further consideration motions which cannot be realized physically it is assumed that the configurations $\{B_t\}$ are similar to one another in the sense that each triple product $[\vec{px}, \vec{py}, \vec{pz}]$, where p, x, y, z are neighbouring points occupied by four particles, is of fixed sign. It will also be supposed that B_r is similar to B_t *and that* $\boldsymbol{\psi}$ *and* $\boldsymbol{\Psi}$ *are twice continuously differentiable jointly in the position and time variables on which they depend.* Thus $\mathscr{B}$ could, in principle, occupy the reference configuration B_r, though it need not actually do so at a time in the interval I. The italicized smoothness requirement implies that curves, surfaces and regions in B_r

are carried by the motion into curves, surfaces and regions in B_t. Such subsets of B_t are occupied by the same particles at all times in I and are consequently referred to as *material curves, surfaces and regions.*

The term *deformation* is employed when the functions $\boldsymbol{\psi}$ and $\boldsymbol{\Psi}$ do not depend upon t and in contexts in which the time dependence is irrelevant. More precisely, a deformation is a (smooth) mapping of a reference configuration into a current (or deformed) configuration. Evidently a motion may be viewed as a one-parameter family of deformations indexed by the time.

Problem 1 A motion of a body $\mathscr{B}$ is said to be *rigid* if the distance between the points occupied by every pair of particles is invariable. Show that for such a motion equations (3) and (4) take the forms

$$\boldsymbol{x} = \boldsymbol{c}(t) + \boldsymbol{Q}(t)\boldsymbol{X}, \quad \boldsymbol{X} = \boldsymbol{Q}^{\mathrm{T}}(t)\{\boldsymbol{x} - \boldsymbol{c}(t)\}, \tag{5}$$

where $\boldsymbol{Q}$ is a proper orthogonal tensor. Use these equations to calculate the velocity and acceleration fields associated with a rigid motion.

Solution. Let $\mathscr{P}, \mathscr{X}, \mathscr{Y}$ be three distinct particles of $\mathscr{B}$ and let $\boldsymbol{P}$, $\boldsymbol{X}$, $\boldsymbol{Y}$ and $\boldsymbol{p}$, $\boldsymbol{x}$, $\boldsymbol{y}$ be the positions of the points occupied by these particles in the reference configuration B_r and the current configuration B_t relative to origins O and o respectively. The rigidity of the motion requires that

$$|\boldsymbol{x} - \boldsymbol{y}| = |\boldsymbol{X} - \boldsymbol{Y}|, \quad |\boldsymbol{x} - \boldsymbol{p}| = |\boldsymbol{X} - \boldsymbol{P}|, \quad |\boldsymbol{y} - \boldsymbol{p}| = |\boldsymbol{Y} - \boldsymbol{P}|,$$

whence

$$(\boldsymbol{x} - \boldsymbol{p}) \cdot (\boldsymbol{y} - \boldsymbol{p}) = (\boldsymbol{X} - \boldsymbol{P}) \cdot (\boldsymbol{Y} - \boldsymbol{P}). \tag{A}$$

Suppose now that $\mathscr{P}$ occupies an interior point of B_r so that if $g = \{\boldsymbol{g}_1, \boldsymbol{g}_2, \boldsymbol{g}_3\}$ is an orthonormal basis of E, the three points with positions $\boldsymbol{P} + l\boldsymbol{g}_i$ $(i = 1, 2, 3)$ relative to O are occupied by particles $\mathscr{Q}_i \in \mathscr{B}$ for a small enough value of the scalar l. If the position relative to o of the point occupied by $\mathscr{Q}_i$ in B_t is $\boldsymbol{p} + l\boldsymbol{e}_i$, it follows from (A) on first replacing $\mathscr{Y}$ by $\mathscr{Q}_i$, then replacing $\mathscr{X}, \mathscr{Y}$ by $\mathscr{Q}_i, \mathscr{Q}_j$, that

$$(\boldsymbol{x} - \boldsymbol{p}) \cdot \boldsymbol{e}_i = (\boldsymbol{X} - \boldsymbol{P}) \cdot \boldsymbol{g}_i \quad \text{and} \quad \boldsymbol{e}_i \cdot \boldsymbol{e}_j = \boldsymbol{g}_i \cdot \boldsymbol{g}_j = \delta_{ij}. \tag{B}$$

$e = \{\boldsymbol{e}_1, \boldsymbol{e}_2, \boldsymbol{e}_3\}$ is therefore a second orthonormal basis of E, and because B_r and B_t are similar configurations, g and e are similar bases. On referring $\boldsymbol{x} - \boldsymbol{p}$ to the basis e we obtain, with the aid of equations (1.14), $(B)_1$ and (1.41),

$$\begin{aligned} \boldsymbol{x} - \boldsymbol{p} &= \{(\boldsymbol{x} - \boldsymbol{p}) . \boldsymbol{e}_p\}\boldsymbol{e}_p = \{(\boldsymbol{X} - \boldsymbol{P}) . \boldsymbol{g}_p\}\boldsymbol{e}_p \\ &= (\boldsymbol{e}_p \otimes \boldsymbol{g}_p)(\boldsymbol{X} - \boldsymbol{P}) = \boldsymbol{Q}(\boldsymbol{X} - \boldsymbol{P}), \end{aligned}$$

where $\boldsymbol{Q} = \boldsymbol{e}_p \otimes \boldsymbol{g}_p$ is a proper orthogonal tensor (see Exercise 1.10, p. 48). Equation $(5)_1$ is now obtained by setting $\boldsymbol{c} = \boldsymbol{p} - \boldsymbol{QP}$ and observing that $\boldsymbol{p}$ and $\boldsymbol{e}_i$, and hence $\boldsymbol{c}$ and $\boldsymbol{Q}$, are time-dependent. The inverse relation $(5)_2$ is derived from $(5)_1$ with the aid of equation (1.75).

The velocity $\boldsymbol{v}$ and the acceleration $\boldsymbol{a}$ are calculated by forming the first and second time derivatives of the spatial position $\boldsymbol{x}$ holding fixed the referential position $\boldsymbol{X}$. Thus, from $(5)_1$,

$$\boldsymbol{v} = \dot{\boldsymbol{c}} + \dot{\boldsymbol{Q}}\boldsymbol{X}, \quad \boldsymbol{a} = \ddot{\boldsymbol{c}} + \ddot{\boldsymbol{Q}}\boldsymbol{X}, \tag{C}$$

a superposed dot denoting differentiation with respect to t. With the use of $(5)_2$, $\boldsymbol{v}$ and $\boldsymbol{a}$ can be expressed in terms of $\boldsymbol{x}$ and t as

$$\boldsymbol{v} = \dot{\boldsymbol{c}} + \dot{\boldsymbol{Q}}\boldsymbol{Q}^{\mathrm{T}}(\boldsymbol{x} - \boldsymbol{c}), \quad \boldsymbol{a} = \ddot{\boldsymbol{c}} + \ddot{\boldsymbol{Q}}\boldsymbol{Q}^{\mathrm{T}}(\boldsymbol{x} - \boldsymbol{c}). \tag{D}$$

Now differentiation with respect to t of the orthogonality relation $\boldsymbol{Q}\boldsymbol{Q}^{\mathrm{T}} = \boldsymbol{I}$ gives $\dot{\boldsymbol{Q}}\boldsymbol{Q}^{\mathrm{T}} = -\boldsymbol{Q}\dot{\boldsymbol{Q}}^{\mathrm{T}} = -(\dot{\boldsymbol{Q}}\boldsymbol{Q}^{\mathrm{T}})^{\mathrm{T}}$, which shows that the tensor $\boldsymbol{W} = \dot{\boldsymbol{Q}}\boldsymbol{Q}^{\mathrm{T}}$ is skew-symmetric. Moreover,

$$\ddot{\boldsymbol{Q}}\boldsymbol{Q}^{\mathrm{T}} = (\dot{\boldsymbol{Q}}\boldsymbol{Q}^{\mathrm{T}})^{\cdot} - \dot{\boldsymbol{Q}}\dot{\boldsymbol{Q}}^{\mathrm{T}} = \dot{\boldsymbol{W}} - (\dot{\boldsymbol{Q}}\boldsymbol{Q}^{\mathrm{T}})(\boldsymbol{Q}\dot{\boldsymbol{Q}}^{\mathrm{T}}) = \dot{\boldsymbol{W}} + \boldsymbol{W}^2.$$

Hence equations (D) can be rewritten as

$$\boldsymbol{v} = \dot{\boldsymbol{c}} + \boldsymbol{W}(\boldsymbol{x} - \boldsymbol{c}), \quad \boldsymbol{a} = \ddot{\boldsymbol{c}} + (\dot{\boldsymbol{W}} + \boldsymbol{W}^2)(\boldsymbol{x} - \boldsymbol{c}), \tag{6}$$

or, in terms of the axial vector $\boldsymbol{w}$ of $\boldsymbol{W}$,

$$\begin{aligned} \boldsymbol{v} = \dot{\boldsymbol{c}} + \boldsymbol{w} \wedge (\boldsymbol{x} - \boldsymbol{c}), \quad \boldsymbol{a} = \ddot{\boldsymbol{c}} &+ \dot{\boldsymbol{w}} \wedge (\boldsymbol{x} - \boldsymbol{c}) \\ &+ \boldsymbol{w} \wedge \{\boldsymbol{w} \wedge (\boldsymbol{x} - \boldsymbol{c})\}. \end{aligned} \tag{7}$$

From equations (7) $\boldsymbol{w}$ is recognized as the *angular velocity* of $\mathscr{B}$, and $\dot{\boldsymbol{c}}$ and $\ddot{\boldsymbol{c}}$ as the velocity and the acceleration of the particle momentarily occupying, at time t, the point with position $\boldsymbol{c}$ relative to o.

2 THE REFERENTIAL AND SPATIAL DESCRIPTIONS

In the development of the basic principles of continuum mechanics a body $\mathscr{B}$ is endowed with various physical properties which are represented by scalar, vector and tensor fields defined either on a reference configuration B_r or on the configurations $\{B_t\}$ constituting a motion of $\mathscr{B}$. In the former case the referential position $\boldsymbol{X}$ and the time t serve as independent variables and the fields are said to be given in the *referential description*.[3] In the latter case the independent variables are the current position $\boldsymbol{x}$ and t, and the term *spatial description*[3] is used. A field which is set in the referential description may be transferred to the spatial description with the aid of the relation (4), and this procedure has already been used in Problem 1 in deriving equations (D) from (C). Passage from the spatial to the referential description is similarly effected by means of equation (3).

Problem 2 The field lines of the velocity, given in the spatial description, are called *streamlines*, and a motion in which the velocity depends only upon current position is said to be *steady*. Show that the streamlines and the particle paths are coincident in a steady motion. Is the converse of this result true?

Solution. We note first that in general the particle paths of a motion are fixed curves in $\mathfrak{E}$, whereas the conformation of streamlines varies with time. In a steady motion, however, the velocity field is the same on each of the configurations $\{B_t\}$ and the streamlines are fixed. The stated property will therefore be established if we can show that, for some $t \in I$, the particle path and the streamline passing through an arbitrary point x in B_t have a common tangent at x. But in any motion a particle path is tangent to a streamline at the point currently occupied by the particle since, by definition, the velocity at that point is directed along both curves.

Consider the motion given by $\boldsymbol{x} = \boldsymbol{X} + \frac{1}{2}kt^2\boldsymbol{p}$, where k is a positive constant and $\boldsymbol{p}$ a fixed unit vector. The velocity is $\boldsymbol{v} = kt\boldsymbol{p}$ and both

[3] The terms *Lagrangian* and *Eulerian*, used in many books and especially in texts dealing specifically with fluid mechanics, are synonymous with *referential* and *spatial* respectively.

the particle paths and the streamlines are straight lines in the direction defined by $\boldsymbol{p}$. But the motion is seen to be unsteady. It is not generally true, therefore, that a motion with coincident particle paths and streamlines is steady.

Rectangular Cartesian systems of *referential* and *spatial coordinates* are set up by adjoining to the repective origins O and o similar orthonormal bases $E = \{\boldsymbol{E}_1, \boldsymbol{E}_2, \boldsymbol{E}_3\}$ and $e = \{\boldsymbol{e}_1, \boldsymbol{e}_2, \boldsymbol{e}_3\}$. Referential coordinates in the system (O, E) are denoted by $X_\alpha (\alpha = 1, 2, 3)$ and spatial coordinates in the system (o, e) by x_i $(i = 1, 2, 3)$. Vectors may be referred to either of the bases E, e and tensors to either basis or to both. Thus a vector field $\boldsymbol{u}$ and a tensor field $\boldsymbol{T}$ have the representations[4]

$$\left.\begin{aligned} &\boldsymbol{u} = u_\pi \boldsymbol{E}_\pi = u_p \boldsymbol{e}_p, \\ &\boldsymbol{T} = T_{\pi\rho} \boldsymbol{E}_\pi \otimes \boldsymbol{E}_\rho = T_{\pi p} \boldsymbol{E}_\pi \otimes \boldsymbol{e}_p = T_{p\pi} \boldsymbol{e}_p \otimes \boldsymbol{E}_\pi = T_{pq} \boldsymbol{e}_p \otimes \boldsymbol{e}_q, \end{aligned}\right\} \tag{8}$$

where the various sets of components are given by

$$\left.\begin{aligned} &u_\alpha = \boldsymbol{u} . \boldsymbol{E}_\alpha, \quad u_i = \boldsymbol{u} . \boldsymbol{e}_i, \\ &T_{\alpha\beta} = \boldsymbol{E}_\alpha . (\boldsymbol{T}\boldsymbol{E}_\beta), \quad T_{\alpha i} = \boldsymbol{E}_\alpha . (\boldsymbol{T}\boldsymbol{e}_i), \quad T_{i\alpha} = \boldsymbol{e}_i . (\boldsymbol{T}\boldsymbol{E}_\alpha), \\ &\qquad\qquad\qquad\qquad\qquad\qquad T_{ij} = \boldsymbol{e}_i . (\boldsymbol{T}\boldsymbol{e}_j). \end{aligned}\right\} \tag{9}$$

Problem 3 The *streakline* through a point p at time t is the curve on which lie the points occupied at that instant by all particles which have occupied p at some earlier time. Obtain the equations representing this curve for the motion given, for $t \geqslant 0$, by

$$x_1 = X_1(1 + kt), \quad x_2 = X_2 + lt, \quad x_3 = X_3, \tag{A}$$

where the referential and spatial coordinates relate to a common rectangular Cartesian system and k, l are positive constants.

Solution. Let $\mathscr{X}$ be a particle occupying at time t a point x on the streakline through p. The coordinates x_i of x are connected to t and the referential coordinates X_α of $\mathscr{X}$ by equations (A). Also, because $\mathscr{X}$ occupies p at some time t' in the interval $[0, t]$,

$$p_1 = X_1(1 + kt'), \quad p_2 = X_2 + lt', \quad p_3 = X_3, \tag{B}$$

p_i being the coordinates of p. A representation of the streakline is

[4] The reader is reminded of the summation convention stated on p. 14.

obtained by eliminating the X_α and t' between equations (B) and (A). Elimination of the X_α gives

$$x_1(1 + kt') = p_1(1 + kt), \quad x_2 - p_2 = l(t - t'), \quad x_3 = p_3,$$

and on eliminating t' we obtain

$$x_1\{x_2 - p_2 - (l/k)(1 + kt)\} = -p_1(l/k)(1 + kt), \quad x_3 = p_3.$$

[The streakline is seen to be an arc of a rectangular hyperbola situated in the plane $x_3 = p_3$. The case in which $p_1 > 0$, $p_2 > 0$ is illustrated in Figure 1. If particles are imagined to be marked with a dye on passing through p the streakline is the curve which would be observed at time t. Note, however, that at this instant the coloured particles are moving along their paths, which are straight lines; only one particle (that occupying the point p) is actually moving along the streakline.]

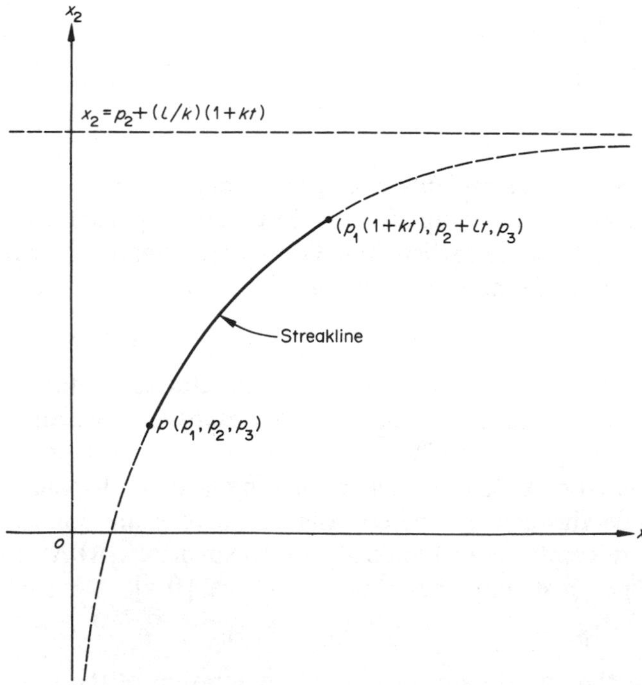

FIGURE 1 Current configuration of a streakline (Problem 3).

Let ϕ be a scalar field specifying some property of a body $\mathscr{B}$. In a motion of $\mathscr{B}$ the *material derivative* of ϕ is the rate of change of ϕ with time at a fixed particle; it is denoted by $\dot{\phi}$. Since the particles of $\mathscr{B}$ are labelled by their positions in a reference configuration, $\dot{\phi}$ is calculated by differentiating ϕ with respect to t holding fixed the referential position $\boldsymbol{X}$. The result is

$$\dot{\phi}(\boldsymbol{X}, t) = \frac{\partial \phi}{\partial t}(\boldsymbol{X}, t) \quad \text{or} \quad \dot{\phi}(\boldsymbol{x}, t) = \frac{\partial \phi}{\partial t}(\boldsymbol{x}, t) + \{(\operatorname{grad} \phi)(\boldsymbol{x}, t)\} \cdot \boldsymbol{v}(\boldsymbol{x}, t) \tag{10}$$

according as ϕ is given in the referential or the spatial description. The corresponding formulae for a vector field $\boldsymbol{u}$ are

$$\dot{\boldsymbol{u}}(\boldsymbol{X}, t) = \frac{\partial \boldsymbol{u}}{\partial t}(\boldsymbol{X}, t), \quad \dot{\boldsymbol{u}}(\boldsymbol{x}, t) = \frac{\partial \boldsymbol{u}}{\partial t}(\boldsymbol{x}, t) + \{(\operatorname{grad} \boldsymbol{u})(\boldsymbol{x}, t)\}\, \boldsymbol{v}(\boldsymbol{x}, t). \tag{11}$$

Replacing $\boldsymbol{u}$ successively by $\boldsymbol{x}$ and $\boldsymbol{v}$ in (11) we obtain

$$\boldsymbol{v} = \dot{\boldsymbol{x}}, \quad \boldsymbol{a} = \dot{\boldsymbol{v}} = \frac{\partial \boldsymbol{v}}{\partial t} + (\operatorname{grad} \boldsymbol{v})\boldsymbol{v} = \ddot{\boldsymbol{x}}, \tag{12}$$

these equations being set in the spatial description.

Problem 4 The equation $\eta(\boldsymbol{x}, t) = 0$ defines a surface S_t in each of the configurations B_t occupied by a moving body $\mathscr{B}$. Prove that S_t is a material surface if and only if $\dot{\eta} = 0$.

Solution. (*a*) η is a scalar field defined on the configurations B_t and if $\dot{\eta} = 0$ the value of η at an arbitrary particle of $\mathscr{B}$ remains fixed during the motion. In particular, those particles for which $\eta = 0$ at some time retain this value for all times, which means that the points of S_t are occupied by the same particles throughout the motion. S_t is accordingly a material surface.

(*b*) Since the points of a material surface are permanently occupied by the same particles such a surface must be specified, in the referential description, by an equation of the form $\zeta(\boldsymbol{X}) = 0$. If S_t is a material surface the function η, when transferred to this description, must therefore depend only upon the referential position $\boldsymbol{X}$. It then follows from equation $(10)_1$ that $\dot{\eta} = 0$.

3 THE DEFORMATION AND VELOCITY GRADIENTS

Extending the notation introduced in Section 1.10, we denote the gradient, divergence and curl operators with regard to position in the reference configuration B_r by Grad, Div and Curl, and the corresponding operators with reference to spatial position in the current configuration B_t by grad, div and curl.

The fundamental kinematic tensor underlying the local analysis of deformation and motion is the *deformation gradient* $\boldsymbol{F}$, defined by

$$\boldsymbol{F}(\boldsymbol{X}, t) = (\text{Grad}\,\boldsymbol{\psi})(\boldsymbol{X}, t), \quad \text{or} \quad \boldsymbol{F} = \text{Grad}\,\boldsymbol{x}. \tag{13}$$

In relation to the systems (O, E) and (o, e) of referential and spatial coordinates,

$$\boldsymbol{F} = F_{p\pi}\boldsymbol{e}_p \otimes \boldsymbol{E}_\pi \quad \text{where } F_{i\alpha} = \frac{\partial x_i}{\partial X_\alpha}. \tag{14}$$

The smoothness conditions imposed in Section 1 (p. 51) ensure the existence of the two sets of partial derivatives $\partial\psi_i/\partial X_\alpha$ and $\partial\Psi_\alpha/\partial x_i$. This means that $\boldsymbol{F}$ has an inverse, defined by

$$\boldsymbol{F}^{-1}(\boldsymbol{x}, t) = (\text{grad}\,\boldsymbol{\Psi})(\boldsymbol{x}, t), \quad \text{or} \quad \boldsymbol{F}^{-1} = \text{grad}\,\boldsymbol{X}, \tag{15}$$

and having the component form

$$\boldsymbol{F}^{-1} = F^{-1}_{\pi p}\boldsymbol{E}_\pi \otimes \boldsymbol{e}_p \quad \text{where } F^{-1}_{\alpha i} = \frac{\partial X_\alpha}{\partial x_i}. \tag{16}$$

Also, since $\boldsymbol{F}$ is invertible,

$$J = \det \boldsymbol{F} \neq 0. \tag{17}$$

Problem 5 If ϕ is a scalar field and $\boldsymbol{u}$ a vector field representing properties of a moving body, show that

$$\text{Grad}\,\phi = \boldsymbol{F}^{\mathrm{T}}\,\text{grad}\,\phi, \quad \text{Grad}\,\boldsymbol{u} = (\text{grad}\,\boldsymbol{u})\boldsymbol{F}. \tag{18}$$

Solution. The stated results can be derived by referring $\boldsymbol{F}$ and the gradients to the coordinate systems (O, E) and (o, e) with the aid of equations (14), (1.92) and $(1.93)_1$. The subsequent manipulations proceed as follows:

$$\operatorname{Grad}\phi - \boldsymbol{F}^{\mathrm{T}}\operatorname{grad}\phi = \frac{\partial\phi}{\partial X_\pi}\boldsymbol{E}_\pi - \left(\frac{\partial x_p}{\partial X_\pi}\boldsymbol{E}_\pi \otimes \boldsymbol{e}_p\right)\left(\frac{\partial\phi}{\partial x_q}\boldsymbol{e}_q\right)$$

$$= \left(\frac{\partial\phi}{\partial X_\pi} - \frac{\partial\phi}{\partial x_q}\frac{\partial x_p}{\partial X_\pi}\delta_{qp}\right)\boldsymbol{E}_\pi = \left(\frac{\partial\phi}{\partial X_\pi} - \frac{\partial\phi}{\partial x_p}\frac{\partial x_p}{\partial X_\pi}\right)\boldsymbol{E}_\pi = \boldsymbol{0}.$$

$$\operatorname{Grad}\boldsymbol{u} - (\operatorname{grad}\boldsymbol{u})\boldsymbol{F} = \frac{\partial u_\pi}{\partial X_\rho}\boldsymbol{E}_\pi \otimes \boldsymbol{E}_\rho - \left(\frac{\partial u_p}{\partial x_q}\boldsymbol{e}_p \otimes \boldsymbol{e}_q\right)\left(\frac{\partial x_r}{\partial X_\rho}\boldsymbol{e}_r \otimes \boldsymbol{E}_\rho\right)$$

$$= \left(\frac{\partial u_\pi}{\partial X_\rho}\boldsymbol{E}_\pi - \frac{\partial u_p}{\partial x_q}\frac{\partial x_r}{\partial X_\rho}\delta_{qr}\boldsymbol{e}_p\right)\otimes \boldsymbol{E}_\rho$$

$$= \left\{\frac{\partial}{\partial X_\rho}(u_\pi \boldsymbol{E}_\pi - u_p \boldsymbol{e}_p)\right\}\otimes \boldsymbol{E}_\rho = \boldsymbol{O}.$$

Problem 6 If $\boldsymbol{u}$ is a vector field and $\boldsymbol{T}$ a tensor field representing properties of a moving body, show that

$$\operatorname{Div}\boldsymbol{u} = J\operatorname{div}(J^{-1}\boldsymbol{F}\boldsymbol{u}), \quad \operatorname{Div}\boldsymbol{T} = J\operatorname{div}(J^{-1}\boldsymbol{F}\boldsymbol{T}). \tag{19}$$

Solution. Equations (19) depend upon the result

$$\frac{\partial}{\partial x_p}(J^{-1}F_{p\alpha}) = 0$$

which we justify by appeal to Problem 1.6 (p. 20). Replacing A and τ by F and X_α in the formula (1.40) we have

$$\frac{\partial J}{\partial X_\alpha} = J\operatorname{tr}\left(\frac{\partial \boldsymbol{F}}{\partial X_\alpha}\boldsymbol{F}^{-1}\right) = J\frac{\partial F_{q\rho}}{\partial X_\alpha}F^{-1}_{\rho q} = J\frac{\partial^2 x_q}{\partial X_\alpha \partial X_\rho}F^{-1}_{\rho q},$$

use being made of equations (17), (1.50) and $(14)_2$. Hence

$$\frac{\partial}{\partial x_p}(J^{-1}F_{p\alpha}) = \frac{\partial}{\partial X_\pi}(J^{-1}F_{p\alpha})\frac{\partial X_\pi}{\partial x_p}$$

$$= \left(J^{-1}\frac{\partial^2 x_p}{\partial X_\alpha \partial X_\pi} - J^{-1}\frac{\partial^2 x_q}{\partial X_\pi \partial X_\rho}F^{-1}_{\rho q}F_{p\alpha}\right)F^{-1}_{\pi p}$$

$$= J^{-1}\left(\frac{\partial^2 x_q}{\partial X_\alpha \partial X_\rho}F^{-1}_{\rho q} - \frac{\partial^2 x_q}{\partial X_\pi \partial X_\rho}\delta_{\pi\alpha}F^{-1}_{\rho q}\right) = 0.$$

We now refer the divergences to the appropriate coordinate systems by means of equations $(1.93)_2$ and (1.94), obtaining

$$\operatorname{Div} \boldsymbol{u} - J \operatorname{div}(J^{-1}\boldsymbol{F}\boldsymbol{u}) = \frac{\partial u_\pi}{\partial X_\pi} - J\frac{\partial}{\partial x_p}(J^{-1}F_{p\pi}u_\pi)$$

$$= \frac{\partial u_\pi}{\partial X_\pi} - F_{p\pi}\frac{\partial u_\pi}{\partial x_p} = \frac{\partial u_\pi}{\partial X_\pi} - \frac{\partial u_\pi}{\partial x_p}\frac{\partial x_p}{\partial X_\pi} = 0,$$

and[5]

$$\operatorname{Div} \boldsymbol{T} - J \operatorname{div}(J^{-1}\boldsymbol{F}\boldsymbol{T}) = \frac{\partial T_{\pi\rho}}{\partial X_\pi}\boldsymbol{E}_\rho - J\frac{\partial}{\partial x_p}(J^{-1}F_{p\pi}T_{\pi q})\boldsymbol{e}_q$$

$$= \frac{\partial T_{\pi\rho}}{\partial X_\pi}\boldsymbol{E}_\rho - F_{p\pi}\frac{\partial T_{\pi q}}{\partial x_p}\boldsymbol{e}_q = \frac{\partial}{\partial X_\pi}(T_{\pi\rho}\boldsymbol{E}_\rho - T_{\pi q}\boldsymbol{e}_q) = \boldsymbol{0}.$$

The remainder of this chapter rests mainly on considerations relating to the situations illustrated in Figures 2, 3 and 4. In Figure 2, C_t and D_t are material curves in B_t, the current configuration of a moving body. These curves intersect at the point x and C_r and D_r, their images in the reference configuration B_r, meet at X. Elementary arcs of C_r and D_r at X, represented by the vectors $\mathrm{d}\boldsymbol{X}$ and $\mathrm{d}\boldsymbol{Y}$, are carried into the elementary arcs of C_t and D_t at x represented by $\mathrm{d}\boldsymbol{x}$ and $\mathrm{d}\boldsymbol{y}$. These vectors are related through the deformation gradient (evaluated at X in the referential description and at x in the spatial description) by

$$\mathrm{d}\boldsymbol{x} = \boldsymbol{F}\mathrm{d}\boldsymbol{X}, \quad \mathrm{d}\boldsymbol{y} = \boldsymbol{F}\,\mathrm{d}\boldsymbol{Y}. \tag{20}$$

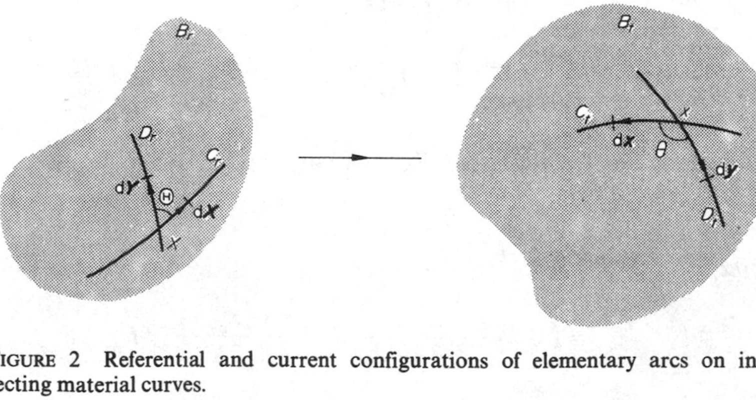

FIGURE 2 Referential and current configurations of elementary arcs on intersecting material curves.

[5] In amplification of the final step, note that $T_{\alpha\rho}\boldsymbol{E}_\rho = T_{\alpha q}\boldsymbol{e}_q = \boldsymbol{T}^{\mathrm{T}}\boldsymbol{E}_\alpha$ from equations $(8)_{3,4}$.

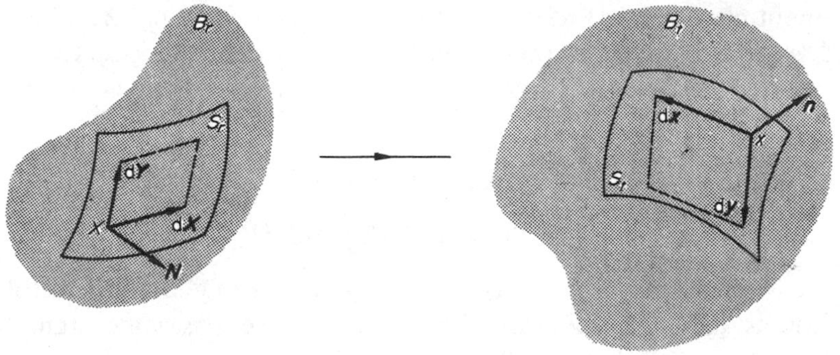

FIGURE 3 Referential and current configurations of an element of a material surface.

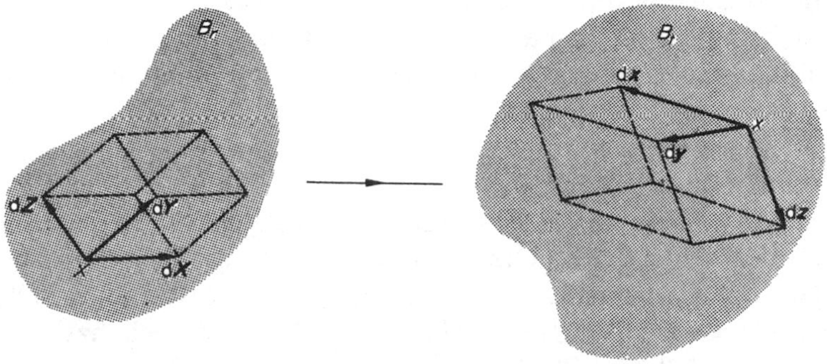

FIGURE 4 Referential and current configurations of an element of a material region.

Figure 3 shows a material surface S_t in B_t with configuration S_r in B_r. An element of area $\mathrm{d}A$ at the point X on S_r, defined by the vectors $\mathrm{d}\boldsymbol{X}$ and $\mathrm{d}\boldsymbol{Y}$, each tangent to S_r, is mapped into the element of area $\mathrm{d}a$ at $x \in S_t$ defined by the vectors $\mathrm{d}\boldsymbol{x}$ and $\mathrm{d}\boldsymbol{y}$ into which $\mathrm{d}\boldsymbol{X}$ and $\mathrm{d}\boldsymbol{Y}$ are carried. Thus $\boldsymbol{N}\mathrm{d}A = \mathrm{d}\boldsymbol{X} \wedge \mathrm{d}\boldsymbol{Y}$ and $\boldsymbol{n}\mathrm{d}a = \mathrm{d}\boldsymbol{x} \wedge \mathrm{d}\boldsymbol{y}$, where $\boldsymbol{N}$ and $\boldsymbol{n}$ are unit vectors normal to S_r and S_t at X and x respectively, and the connection between these directed areas, established via (20) with the aid of equations (1.38) and (1.39), is

$$\boldsymbol{n}\,\mathrm{d}a = (\boldsymbol{F}\,\mathrm{d}\boldsymbol{X}) \wedge (\boldsymbol{F}\,\mathrm{d}\boldsymbol{Y}) = \boldsymbol{F}^*(\mathrm{d}\boldsymbol{X} \wedge \mathrm{d}\boldsymbol{Y}) = J(\boldsymbol{F}^{-1})^{\mathrm{T}}\boldsymbol{N}\,\mathrm{d}A. \quad (21)$$

In Figure 4, x is an interior point of a material region and the particle currently situated there occupies the point X in B_r. An element of volume $\mathrm{d}V$ at X, defined by the three vectors $\mathrm{d}\boldsymbol{X}$, $\mathrm{d}\boldsymbol{Y}$ and $\mathrm{d}\boldsymbol{Z}$, is carried into the element of volume $\mathrm{d}v$ at x defined by the vectors $\mathrm{d}\boldsymbol{x} = \boldsymbol{F}\mathrm{d}\boldsymbol{X}$, $\mathrm{d}\boldsymbol{y} = \boldsymbol{F}\mathrm{d}\boldsymbol{Y}$ and $\mathrm{d}\boldsymbol{z} = \boldsymbol{F}\mathrm{d}\boldsymbol{Z}$. Hence, by appeal to the definition (1.30),

$$\mathrm{d}v = \pm[\mathrm{d}\boldsymbol{x}, \mathrm{d}\boldsymbol{y}, \mathrm{d}\boldsymbol{z}] = \pm[\boldsymbol{F}\mathrm{d}\boldsymbol{X}, \boldsymbol{F}\mathrm{d}\boldsymbol{Y}, \boldsymbol{F}\mathrm{d}\boldsymbol{Z}]$$
$$= \pm \det \boldsymbol{F}[\mathrm{d}\boldsymbol{X}, \mathrm{d}\boldsymbol{Y}, \mathrm{d}\boldsymbol{Z}] = J\,\mathrm{d}V, \tag{22}$$

the assumed similarity of B_r and B_t guaranteeing that the triple products $[\mathrm{d}\boldsymbol{X}, \mathrm{d}\boldsymbol{Y}, \mathrm{d}\boldsymbol{Z}]$ and $[\mathrm{d}\boldsymbol{x}, \mathrm{d}\boldsymbol{y}, \mathrm{d}\boldsymbol{z}]$ have the same sign. It follows from (22) that $J > 0$.

Problem 7 Obtain conditions on the deformation gradient ensuring that, at a given particle,

(i) no extension occurs in a specified direction,
(ii) the angle between a pair of specified directions is unchanged,
(iii) no change of area takes place in the plane orthogonal to a specified direction,
(iv) no change of volume occurs.

Solution. (i) With reference to Figure 2, let the unit tangent vectors to C_r at X and to C_t at x be $\boldsymbol{L}$ and $\boldsymbol{l}$ respectively. Then $\mathrm{d}\boldsymbol{X} = \boldsymbol{L}\mathrm{d}S$ and $\mathrm{d}\boldsymbol{x} = \boldsymbol{l}\mathrm{d}s$, where $\mathrm{d}S$ and $\mathrm{d}s$ are the arc lengths, and equation $(20)_1$ gives $\boldsymbol{l}\mathrm{d}s = \boldsymbol{F}\boldsymbol{L}\mathrm{d}S$, $\boldsymbol{L}\mathrm{d}S = \boldsymbol{F}^{-1}\boldsymbol{l}\mathrm{d}s$. Since $\boldsymbol{L}$ and $\boldsymbol{l}$ each have unit norm, there follow, with the use of (1.24), the relations

$$(\mathrm{d}s/\mathrm{d}S)^2 = \boldsymbol{L}\,.\,(\boldsymbol{C}\boldsymbol{L}) = \{\boldsymbol{l}\,.\,(\boldsymbol{B}^{-1}\boldsymbol{l})\}^{-1}, \tag{23}$$

where

$$\boldsymbol{C} = \boldsymbol{F}^{\mathrm{T}}\boldsymbol{F}, \quad \boldsymbol{B} = \boldsymbol{F}\boldsymbol{F}^{\mathrm{T}}. \tag{24}$$

If no extension takes place in the direction defined by $\boldsymbol{L}$ at X and by $\boldsymbol{l}$ at x, $\mathrm{d}s = \mathrm{d}S$ and the resulting condition on the deformation gradient $\boldsymbol{F}$, supplied by (23), can be stated in the alternative forms

$$\boldsymbol{L}\,.\,(\boldsymbol{C}\boldsymbol{L}) = 1, \quad \boldsymbol{l}\,.\,(\boldsymbol{B}^{-1}\boldsymbol{l}) = 1. \tag{25}$$

(ii) Again we refer to Figure 2. Let $\boldsymbol{M}$, $\mathrm{d}U$ and $\boldsymbol{m}$, $\mathrm{d}u$ be the unit tangent vectors and arc lengths on D_r at X and on D_t at x respectively, so that $\mathrm{d}\boldsymbol{Y} = \boldsymbol{M}\mathrm{d}U$, $\mathrm{d}\boldsymbol{y} = \boldsymbol{m}\mathrm{d}u$. Then if Θ and θ are the indicated angles of intersection, $\cos\Theta = \boldsymbol{L}.\boldsymbol{M}$ and $\cos\theta = \boldsymbol{l}\,.\,\boldsymbol{m}$. From equations (20),

$$\mathrm{d}\boldsymbol{x}\,.\,\mathrm{d}\boldsymbol{y} = (\boldsymbol{F}\mathrm{d}\boldsymbol{X})\,.\,(\boldsymbol{F}\mathrm{d}\boldsymbol{Y}) = \mathrm{d}\boldsymbol{X}\,.\,(\boldsymbol{C}\mathrm{d}\boldsymbol{Y}), \quad \mathrm{d}\boldsymbol{X}\,.\,\mathrm{d}\boldsymbol{Y} = \mathrm{d}\boldsymbol{x}\,.\,(\boldsymbol{B}^{-1}\mathrm{d}\boldsymbol{y}).$$

Thus

$$\left.\begin{aligned} \cos\theta = \boldsymbol{l}\,.\,\boldsymbol{m} &= \{\boldsymbol{L}\,.\,(\boldsymbol{C}\boldsymbol{M})\}\frac{\mathrm{d}S}{\mathrm{d}s}\frac{\mathrm{d}U}{\mathrm{d}u}, \\ \cos\Theta = \boldsymbol{L}\,.\,\boldsymbol{M} &= \{\boldsymbol{l}\,.\,(\boldsymbol{B}^{-1}\boldsymbol{m})\}\frac{\mathrm{d}s}{\mathrm{d}S}\frac{\mathrm{d}u}{\mathrm{d}U}. \end{aligned}\right\} \tag{26}$$

If the angle between the directions defined by $\boldsymbol{L}$ and $\boldsymbol{M}$ at X and by $\boldsymbol{l}$ and $\boldsymbol{m}$ at x is unchanged, $\theta = \Theta$ and, with the use of (23), equations (26) yield the relations

$$\boldsymbol{L}\,.\,(\boldsymbol{C}\boldsymbol{M}) = (\boldsymbol{L}\,.\,\boldsymbol{M})\,[\{\boldsymbol{L}\,.\,(\boldsymbol{C}\boldsymbol{L})\}\,\{\boldsymbol{M}\,.\,(\boldsymbol{C}\boldsymbol{M})\}]^{1/2},$$

$$\boldsymbol{l}\,.\,(\boldsymbol{B}^{-1}\boldsymbol{m}) = (\boldsymbol{l}\,.\,\boldsymbol{m})\,[\{\boldsymbol{l}\,.\,(\boldsymbol{B}^{-1}\boldsymbol{l})\}\,\{\boldsymbol{m}\,.\,(\boldsymbol{B}^{-1}\boldsymbol{m})\}]^{1/2}.$$

These are the referential and spatial forms of the required condition.

(iii) By using the fact that the vectors $\boldsymbol{N}$ and $\boldsymbol{n}$ are of unit norm, we can derive from equations (21) a pair of relations analogous to (23), namely

$$(\mathrm{d}a/\mathrm{d}A)^2 = J^2\boldsymbol{N}\,.\,(\boldsymbol{C}^{-1}\boldsymbol{N}) = J^2\{\boldsymbol{n}\,.\,(\boldsymbol{B}\boldsymbol{n})\}^{-1}. \tag{27}$$

If, in the motion depicted in Figure 3, no change of area occurs as the chosen particle moves from X to x, $\mathrm{d}a = \mathrm{d}A$ and the restriction

$$\boldsymbol{N}\,.\,(\boldsymbol{C}^{-1}\boldsymbol{N}) = J^{-2}, \quad \text{or} \quad \boldsymbol{n}\,.\,(\boldsymbol{B}\boldsymbol{n}) = J^2,$$

applies to $\boldsymbol{F}$.

(iv) In this case the volume elements shown in Figure 4 have equal content so that $\mathrm{d}V = \mathrm{d}v$ and, from equation (22),

$$J = 1. \tag{28}$$

A deformation for which condition (iv) of Problem 7 applies to every particle is said to be *isochoric*. Equation (28), holding on each configuration in $\{B_t\}$, is a necessary and sufficient condition for a motion to have this property.

Problem 8 In its reference configuration a body contains a spherical cavity with centre o and radius A, filled with explosive. The explosive is detonated at $t = 0$ and produces a spherically symmetric motion of the body given by

$$\boldsymbol{x} = (r/R)\boldsymbol{X}, \quad r = f(R, t), \quad \text{where} \quad R = |\boldsymbol{X}|, r = |\boldsymbol{x}|,$$

and the referential and current positions, $\boldsymbol{X}$ and $\boldsymbol{x}$, are both taken relative to o as origin. If the motion is isochoric and the cavity radius at time t is a, show that

$$f(R, t) = (R^3 + a^3 - A^3)^{1/3}, \tag{A}$$

and determine the velocity and acceleration in the spatial description.

Solution. In the given motion the particle paths are segments of straight lines through o, while R and r are respectively the distances from o of the points occupied initially and currently by a typical particle. Spheres with centre o and radius greater than or equal to a are therefore material surfaces and since the motion is isochoric the volume of the region bounded externally by the material sphere of radius r and internally by the cavity surface is invariable. At $t = 0$ the radii of this material region are R and A. Hence $\frac{4}{3}\pi(r^3 - a^3) = \frac{4}{3}\pi(R^3 - A^3)$, which leads directly to (A).

[It is instructive also to follow through the derivation of (A) which starts with the calculation of the deformation gradient. Using Exercise 1.14 (i) (p. 49) and noticing that $\operatorname{Grad} R = R^{-1}\boldsymbol{X}, \operatorname{Grad} \boldsymbol{X} = \boldsymbol{I}$, we have

$$\boldsymbol{F} = \operatorname{Grad} \boldsymbol{x} = \left(\frac{\partial f}{\partial R} - \frac{f}{R}\right) R^{-2}\boldsymbol{X} \otimes \boldsymbol{X} + \frac{f}{R}\boldsymbol{I}.$$

With the help of equation (1.44) this can be rewritten as

$$\boldsymbol{F} = \frac{\partial f}{\partial R}\hat{\boldsymbol{r}} \otimes \hat{\boldsymbol{r}} + \frac{f}{R}(\hat{\boldsymbol{\theta}} \otimes \hat{\boldsymbol{\theta}} + \hat{\boldsymbol{\phi}} \otimes \hat{\boldsymbol{\phi}}),$$

where $\hat{\boldsymbol{r}} = R^{-1}\boldsymbol{X}$ and $\hat{\boldsymbol{\theta}}, \hat{\boldsymbol{\phi}}$ are unit vectors forming with $\hat{\boldsymbol{r}}$ an orthonormal set. We see that $\boldsymbol{F}$ is a symmetric tensor with proper numbers $\partial f/\partial R, f/R, f/R$ and principal axes defined by $\hat{\boldsymbol{r}}, \hat{\boldsymbol{\theta}}, \hat{\boldsymbol{\phi}}$. J, the determinant of $\boldsymbol{F}$, is the product of these proper numbers, so the condition

(28) for isochoric motion gives rise to the first-order differential equation

$$f^2 \frac{\partial f}{\partial R} = R^2 \tag{B}$$

for f. On the cavity surface, $R = A$ and $r = a$. Hence $f(A, t) = a$, and on integrating equation (B) subject to this condition we again reach the result (A).]

In the referential description the velocity and acceleration are given by

$$\boldsymbol{v} = \frac{\partial f}{\partial t} R^{-1}\mathbf{X} = \frac{\partial f}{\partial t}\hat{\boldsymbol{r}}, \quad \boldsymbol{a} = \frac{\partial^2 f}{\partial t^2} R^{-1}\mathbf{X} = \frac{\partial^2 f}{\partial t^2}\hat{\boldsymbol{r}},$$

and, as might have been anticipated, both of these fields are directed radially to o, the centre of the spherically symmetric motion. From (A),

$$\frac{\partial f}{\partial t} = f^{-2}a^2\dot{a}, \quad \frac{\partial^2 f}{\partial t^2} = f^{-2}(a^2\ddot{a} + 2a\dot{a}^2) - 2f^{-3}\frac{\partial f}{\partial t}a^2\dot{a}$$

$$= f^{-2}(a^2\ddot{a} + 2a\dot{a}^2) - 2f^{-5}a^4\dot{a}^2,$$

where $\dot{a} = \mathrm{d}a/\mathrm{d}t$, $\ddot{a} = \mathrm{d}^2 a/\mathrm{d}t^2$. The transfer of these results to the spatial description requires only the replacement of f by r, whence

$$\boldsymbol{v} = a^2\dot{a}r^{-2}\hat{\boldsymbol{r}}, \quad \boldsymbol{a} = \{(a^2\ddot{a} + 2a\dot{a}^2)r^{-2} - 2a^4\dot{a}^2 r^{-5}\}\hat{\boldsymbol{r}}.$$

The *velocity gradient* $\boldsymbol{L}$ is defined by

$$\boldsymbol{L} = \operatorname{grad} \boldsymbol{v} = \operatorname{grad} \dot{\boldsymbol{x}}. \tag{29}$$

Equation $(18)_2$, with $\boldsymbol{u} = \dot{\boldsymbol{x}}$, gives $\operatorname{Grad} \dot{\boldsymbol{x}} = \boldsymbol{LF}$ and since, in the referential description, the gradient and the material derivative commute, there follow from (29), with the use of the definition (13), the relations

$$\dot{\boldsymbol{F}} = \boldsymbol{LF}, \quad \boldsymbol{L} = \dot{\boldsymbol{F}}\boldsymbol{F}^{-1} = -\boldsymbol{F}(\boldsymbol{F}^{-1})^{\cdot} \tag{30}$$

between the velocity and deformation gradients. On writing $\boldsymbol{F}$ and t in place of $\boldsymbol{A}$ and τ in equation (1.40) and using $(30)_2$ we find that

$$\dot{J} = J \operatorname{tr} \boldsymbol{L} = J \operatorname{div} \boldsymbol{v}. \tag{31}$$

In the case of an isochoric motion, $J = 1$ and (31) reduces to

$$\operatorname{tr} \boldsymbol{L} = \operatorname{div} \boldsymbol{v} = 0. \tag{32}$$

Problem 9 In connection with the motion illustrated in Figure 2, determine the material derivatives of the quantities $\mathrm{d}\boldsymbol{x}$, $\boldsymbol{l}$ and $\mathrm{d}s$ relating to an elementrary arc of the material curve C_t. Find also the material derivative of θ, the angle of intersection of C_t and D_t.

Solution. From equations $(20)_1$ and $(30)_1$,

$$(\mathrm{d}\boldsymbol{x})^{\cdot} = \dot{\boldsymbol{F}}\mathrm{d}\boldsymbol{X} = \boldsymbol{LF}\mathrm{d}\boldsymbol{X} = \boldsymbol{L}\,\mathrm{d}\boldsymbol{x},$$

or, replacing $\mathrm{d}\boldsymbol{x}$ by $\boldsymbol{l}\mathrm{d}s$,

$$\boldsymbol{l}(\mathrm{d}s)^{\cdot} + \dot{\boldsymbol{l}}\,\mathrm{d}s = \boldsymbol{Ll}\,\mathrm{d}s.$$

Since $\boldsymbol{l}$ is a unit vector, $\boldsymbol{l}\,.\,\boldsymbol{l} = 1$ and $\boldsymbol{l}\,.\,\dot{\boldsymbol{l}} = 0$. Hence

$$(\mathrm{d}s)^{\cdot} = \boldsymbol{l}\,.\,(\boldsymbol{Ll})\,\mathrm{d}s \quad \text{and} \quad \dot{\boldsymbol{l}} = \boldsymbol{Ll} - \{\boldsymbol{l}\,.\,(\boldsymbol{Ll})\}\boldsymbol{l}. \tag{33}$$

Material differentiation of equation $(26)_1$ yields

$$-\sin\theta\dot{\theta} = \dot{\boldsymbol{l}}\,.\,\boldsymbol{m} + \boldsymbol{l}\,.\,\dot{\boldsymbol{m}}.$$

On using $(33)_2$ and the analogous expression for $\dot{\boldsymbol{m}}$, and setting $\sin\theta = |\boldsymbol{l} \wedge \boldsymbol{m}|$, we arrive at the following formula for $\dot{\theta}$:

$$\dot{\theta} = |\boldsymbol{l} \wedge \boldsymbol{m}|^{-1}[\{\boldsymbol{l}\,.\,(\boldsymbol{Ll}) + \boldsymbol{m}\,.\,(\boldsymbol{Lm})\}\,(\boldsymbol{l}\,.\,\boldsymbol{m}) - \boldsymbol{l}\,.\,\{(\boldsymbol{L} + \boldsymbol{L}^T)\boldsymbol{m}\}]. \tag{34}$$

In conjunction with (30) and (31), equations (20) to (22) lead to a set of *transport formulae* for the rates of change of integrals over material curves, surfaces and regions. Let C_t, S_t and R_t denote in turn a material curve, a material surface and a material region in the current configuration B_t of a body $\mathscr{B}$, and let ϕ be a continuously differentiable scalar field representing some property of $\mathscr{B}$. Then

$$\frac{\mathrm{d}}{\mathrm{d}t}\int_{C_t} \phi\,\mathrm{d}\boldsymbol{x} = \int_{C_t} (\dot{\phi}\,\mathrm{d}\boldsymbol{x} + \phi\boldsymbol{L}\,\mathrm{d}\boldsymbol{x}), \tag{35}$$

$$\frac{\mathrm{d}}{\mathrm{d}t}\int_{S_t} \phi\boldsymbol{n}\,\mathrm{d}a = \int_{S_t} \{(\dot{\phi} + \phi\operatorname{tr}\boldsymbol{L})\boldsymbol{n} - \phi\boldsymbol{L}^T\boldsymbol{n}\}\,\mathrm{d}a, \tag{36}$$

$$\frac{\mathrm{d}}{\mathrm{d}t}\int_{R_t} \phi\,\mathrm{d}v = \int_{R_t} (\dot{\phi} + \phi\operatorname{tr}\boldsymbol{L})\,\mathrm{d}v, \tag{37}$$

where $\boldsymbol{n}$ is a unit normal vector on S_t. The corresponding results for a continuously differentiable vector field $\boldsymbol{u}$ are

$$\frac{\mathrm{d}}{\mathrm{d}t}\int_{C_t} \boldsymbol{u} \,.\, \mathrm{d}\boldsymbol{x} = \int_{C_t} (\dot{\boldsymbol{u}} + \boldsymbol{L}^{\mathrm{T}}\boldsymbol{u}) \,.\, \mathrm{d}\boldsymbol{x}, \tag{38}$$

$$\frac{\mathrm{d}}{\mathrm{d}t}\int_{S_t} \boldsymbol{u} \,.\, \boldsymbol{n}\, \mathrm{d}a = \int_{S_t} (\dot{\boldsymbol{u}} + \boldsymbol{u} \operatorname{tr} \boldsymbol{L} - \boldsymbol{L}\boldsymbol{u}) \,.\, \boldsymbol{n}\, \mathrm{d}a, \tag{39}$$

$$\frac{\mathrm{d}}{\mathrm{d}t}\int_{R_t} \boldsymbol{u}\, \mathrm{d}v = \int_{R_t} (\dot{\boldsymbol{u}} + \boldsymbol{u} \operatorname{tr} \boldsymbol{L})\, \mathrm{d}v. \tag{40}$$

Each of equations (35) to (40) is established by first transferring the expression on the left to the referential description. The range of integration is then a fixed curve, surface or region in the reference configuration B_r and the differentiation with respect to t can be performed under the integral sign. The expression on the right is finally obtained on reverting to the spatial description. The steps in this procedure are given below in the case of equation (39): the reader should write out for himself derivations of the other formulae.

$$\frac{\mathrm{d}}{\mathrm{d}t}\int_{S_t} \boldsymbol{u} \,.\, \boldsymbol{n}\, \mathrm{d}a = \int_{S_r} \frac{\partial}{\partial t}[\boldsymbol{u} \,.\, \{J(\boldsymbol{F}^{-1})^{\mathrm{T}}\boldsymbol{N}\, \mathrm{d}A\}] = \int_{S_r} \frac{\partial}{\partial t}(J\boldsymbol{F}^{-1}\boldsymbol{u}) \,.\, \boldsymbol{N}\, \mathrm{d}A$$

$$= \int_{S_r} \{J\boldsymbol{F}^{-1}\dot{\boldsymbol{u}} + (J \operatorname{tr} \boldsymbol{L})\boldsymbol{F}^{-1}\boldsymbol{u} + J(-\boldsymbol{F}^{-1}\boldsymbol{L})\boldsymbol{u}\} \,.\, \boldsymbol{N}\, \mathrm{d}A$$

$$= \int_{S_r} (\dot{\boldsymbol{u}} + \boldsymbol{u} \operatorname{tr} \boldsymbol{L} - \boldsymbol{L}\boldsymbol{u}) \,.\, \{J(\boldsymbol{F}^{-1})^{\mathrm{T}}\boldsymbol{N}\, \mathrm{d}A\}$$

$$= \int_{S_t} (\dot{\boldsymbol{u}} + \boldsymbol{u} \operatorname{tr} \boldsymbol{L} - \boldsymbol{L}\boldsymbol{u}) \,.\, \boldsymbol{n}\, \mathrm{d}a.$$

The successive steps in this calculation are justified by equations (21), (1.24), (31) and $(30)_3$, (1.24) and (21).

4 STRETCH AND ROTATION

Equations (22) to (24), (26) and (27) indicate that the deformation gradient $\boldsymbol{F}$ compares the sizes and shapes of material elements in the current configuration B_t of a body with the corresponding properties in a reference configuration B_r. We now make a more detailed examination of the geometrical implications of the deformation gradient by utilizing the polar decomposition theorem and the associated results given in Section 1.8. The right and left polar decompositions of $\boldsymbol{F}$ have the respective forms

$$\boldsymbol{F} = \boldsymbol{RU} \quad \text{and} \quad \boldsymbol{F} = \boldsymbol{VR}, \tag{41}$$

where $\boldsymbol{U}$ and $\boldsymbol{V}$ are positive definite symmetric tensors and, since $J > 0$, $\boldsymbol{R}$ is a *proper* orthogonal tensor.

We consider the elementary vector $\mathrm{d}\boldsymbol{x}$, shown in Figure 2, which represents the separation of the points occupied in B_t by a pair of 'nearby' particles. Problem 1.17 (p. 36) tells us first that the actions on such a separation vector of the positive definite symmetric tensors $\boldsymbol{U}$ and $\boldsymbol{V}$ are each equivalent to a triaxial stretch, in directions defined by the appropriate proper vectors and by amounts equal to the proper numbers λ_i common to $\boldsymbol{U}$ and $\boldsymbol{V}$, and secondly that the action of $\boldsymbol{R}$ on $\mathrm{d}\boldsymbol{x}$ can be regarded as a rotation. Further, according to equation $(1.81)_2$, the directions specified by the proper vectors $\boldsymbol{q}_i$ of $\boldsymbol{V}$ are obtained by applying the rotation associated with $\boldsymbol{R}$ to the directions represented by the proper vectors $\boldsymbol{p}_i$ of $\boldsymbol{U}$. From these facts there emerges the following interpretation of equations (41). *The deformation undergone by the body in the locality of a given particle can be effected by first imposing the stretches λ_i in the directions in B_r defined by the orthonormal vectors $\boldsymbol{p}_i$ and then carrying out the rigid rotation given by $\boldsymbol{R}$, or by performing these operations in the reverse order, the directions of stretch then being specified, in B_t, by the orthonormal vectors $\boldsymbol{q}_i$.* Accordingly, $\boldsymbol{R}$ is called the *rotation tensor*, $\boldsymbol{U}$ and $\boldsymbol{V}$ the *right and left stretch tensors*, and λ_i the *principal stretches*; the principal axes associated with the vectors $\boldsymbol{p}_i$ and $\boldsymbol{q}_i$ are termed the *referential* and *current stretch* axes respectively.

On substituting the polar decompositions (41) into the definitions (24) we find that the tensors $\boldsymbol{C}$ and $\boldsymbol{B}$ are simply the squares of the stretch tensors: $\boldsymbol{C} = \boldsymbol{U}^2$, $\boldsymbol{B} = \boldsymbol{V}^2$. $\boldsymbol{C}$ and $\boldsymbol{B}$ are known as the *right and left Cauchy–Green strain tensors.* The referential and current stretch axes coincide with the principal axes of $\boldsymbol{C}$ and $\boldsymbol{B}$ respectively, and the common proper numbers of $\boldsymbol{C}$ and $\boldsymbol{B}$ are the squares of the principal stretches.

Problem 7 (p. 62) shows that the Cauchy–Green strain tensors arise naturally in the calculation of the local changes of length, angle and area engendered by a motion. From the relations

$$\boldsymbol{L}.(\boldsymbol{CL}) = (\mathrm{d}s/\mathrm{d}S)^2, \quad \boldsymbol{L}.(\boldsymbol{CM}) = \cos\theta\,(\mathrm{d}s/\mathrm{d}S)\,(\mathrm{d}u/\mathrm{d}U),$$

given by $(23)_1$ and $(26)_1$ and referring to the pair of intersecting material arcs shown in Figure 2, the following geometrical meanings can be attached to the components $C_{\alpha\beta}$ of the right Cauchy–Green strain tensor relative to the orthonormal basis $\boldsymbol{E}$ and evaluated at the point X in B_r. $C_{\alpha\alpha}$ is the square of the stretch undergone by a material line element which, in B_r, is situated at X and aligned with the direction defined by the base vector $\boldsymbol{E}_\alpha$. $C_{\alpha\beta}$ $(\alpha \neq \beta)$ relates to a pair of material line elements which, in B_r, meet at X and are in the orthogonal directions specified by $\boldsymbol{E}_\alpha$ and $\boldsymbol{E}_\beta$, being the product of the stretches suffered by these elements and the cosine of the angle between them in B_t. The components B_{ij}^{-1} of the inverse left Cauchy–Green strain tensor relative to the basis $\boldsymbol{e}$ can be interpreted in a similar way, the roles of the configurations B_r and B_t now being interchanged.

The next two problems exemplify some basic features of the analysis of deformation in the case of a body occupying permanently a configuration B. A deformation in which $\boldsymbol{F}$ has the same value at every point in the current configuration is said to be *homogeneous*. Such deformations are of major significance in the theory of elasticity.[6]

Problem 10 The homogeneous deformation for which

$$\boldsymbol{F} = \boldsymbol{I} + \gamma \boldsymbol{l} \otimes \boldsymbol{m}, \tag{42}$$

where γ is a positive constant and $\boldsymbol{l}$, $\boldsymbol{m}$ are orthogonal unit vectors, is called a *simple shear*. For this deformation determine the principal stretches, the referential and current stretch axes and the rotation tensor.

Solution. Let $\boldsymbol{n}$ be a unit vector forming with $\boldsymbol{l}$ and $\boldsymbol{m}$ an orthonormal set. The configuration in B of a material region which, in B_r, is a parallelepiped with its faces orthogonal to the directions represented by $\boldsymbol{l}$, $\boldsymbol{m}$ and $\boldsymbol{n}$ is shown in Figure 5: $\boldsymbol{l}$ is said to define the *direction of shear* and the planes orthogonal to $\boldsymbol{m}$ and $\boldsymbol{n}$ are called the *glide planes* and the *planes of shear* respectively.

[6] See Sections 4.2 and 4.8 below.

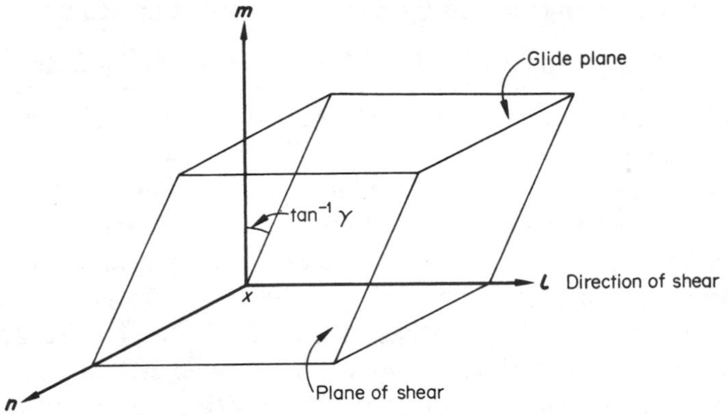

FIGURE 5 Deformed configuration of an initially cuboidal material region in simple shear (Problem 10).

From equation (42),

$$[\boldsymbol{Fl}, \boldsymbol{Fm}, \boldsymbol{Fn}] = [\boldsymbol{l}, \boldsymbol{m} + \gamma \boldsymbol{l}, \boldsymbol{n}] = [\boldsymbol{l}, \boldsymbol{m}, \boldsymbol{n}].$$

Hence $\det \boldsymbol{F} = 1$ and the deformation is isochoric. Also, bearing in mind that $\boldsymbol{I} = \boldsymbol{l} \otimes \boldsymbol{l} + \boldsymbol{m} \otimes \boldsymbol{m} + \boldsymbol{n} \otimes \boldsymbol{n}$,

$$\left.\begin{aligned} \boldsymbol{C} = \boldsymbol{F}^{\mathrm{T}}\boldsymbol{F} &= \boldsymbol{l} \otimes \boldsymbol{l} + (1 + \gamma^2)\boldsymbol{m} \otimes \boldsymbol{m} + \boldsymbol{n} \otimes \boldsymbol{n} \\ &\qquad + \gamma(\boldsymbol{l} \otimes \boldsymbol{m} + \boldsymbol{m} \otimes \boldsymbol{l}), \\ \boldsymbol{B} = \boldsymbol{F}\boldsymbol{F}^{\mathrm{T}} &= (1 + \gamma^2)\boldsymbol{l} \otimes \boldsymbol{l} + \boldsymbol{m} \otimes \boldsymbol{m} + \boldsymbol{n} \otimes \boldsymbol{n} \\ &\qquad + \gamma(\boldsymbol{l} \otimes \boldsymbol{m} + \boldsymbol{m} \otimes \boldsymbol{l}). \end{aligned}\right\} \quad \text{(A)}$$

Now $\boldsymbol{C}$ and $\boldsymbol{B}$ have the spectral representations

$$\boldsymbol{C} = \sum_{r=1}^{3} \lambda_r^2 \boldsymbol{p}_r \otimes \boldsymbol{p}_r, \quad \boldsymbol{B} = \sum_{r=1}^{3} \lambda_r^2 \boldsymbol{q}_r \otimes \boldsymbol{q}_r,$$

the orthonormal triplets $\boldsymbol{p}_i$ and $\boldsymbol{q}_i$ specifying the referential and current stretch axes respectively. From (A), $\boldsymbol{Cn} = \boldsymbol{Bn} = \boldsymbol{n}$, which proves first that one of the principal stretches is unity, and secondly that the associated stretch axis is in the direction defined by $\boldsymbol{n}$ in both B_r and B: in short, the deformation is 'two-dimensional' with respect to the planes of shear. Additionally, $\lambda_1\lambda_2\lambda_3 = \det \boldsymbol{F} = 1$.

The spectral representations can therefore be rewritten as

$$\left.\begin{aligned} \boldsymbol{C} &= \alpha^2 \boldsymbol{p}_1 \otimes \boldsymbol{p}_1 + \alpha^{-2} \boldsymbol{p}_2 \otimes \boldsymbol{p}_2 + \boldsymbol{n} \otimes \boldsymbol{n}, \\ \boldsymbol{B} &= \alpha^2 \boldsymbol{q}_1 \otimes \boldsymbol{q}_1 + \alpha^{-2} \boldsymbol{q}_2 \otimes \boldsymbol{q}_2 + \boldsymbol{n} \otimes \boldsymbol{n}, \end{aligned}\right\} \qquad \text{(B)}$$

where $\alpha = \lambda_1 = \lambda_2^{-1}$.

Since $\boldsymbol{p}_1$, $\boldsymbol{p}_2$ and $\boldsymbol{q}_1$, $\boldsymbol{q}_2$ are orthogonal pairs of unit vectors all orthogonal to $\boldsymbol{n}$ they can be expressed as linear combinations of $\boldsymbol{l}$ and $\boldsymbol{m}$ as follows:

$$\boldsymbol{p}_1 = \cos\Theta\,\boldsymbol{l} + \sin\Theta\,\boldsymbol{m}, \quad \boldsymbol{p}_2 = -\sin\Theta\,\boldsymbol{l} + \cos\Theta\,\boldsymbol{m} \qquad (0 < \Theta < \tfrac{1}{2}\pi),$$

$$\boldsymbol{q}_1 = \cos\theta\,\boldsymbol{l} + \sin\theta\,\boldsymbol{m}, \quad \boldsymbol{q}_2 = -\sin\theta\,\boldsymbol{l} + \cos\theta\,\boldsymbol{m} \qquad (0 < \theta < \tfrac{1}{2}\pi).$$

On entering these expressions into (B) and identifying the resulting equations for $\boldsymbol{C}$ and $\boldsymbol{B}$ with (A), we obtain the relations

$$\alpha^2\cos^2\Theta + \alpha^{-2}\sin^2\Theta = 1, \quad \alpha^2\sin^2\Theta + \alpha^{-2}\cos^2\Theta = 1 + \gamma^2,$$
$$(\alpha^2 - \alpha^{-2})\sin\Theta\cos\Theta = \gamma,$$
$$\alpha^2\cos^2\theta + \alpha^{-2}\sin^2\theta = 1 + \gamma^2, \quad \alpha^2\sin^2\theta + \alpha^{-2}\cos^2\theta = 1,$$
$$(\alpha^2 - \alpha^{-2})\sin\theta\cos\theta = \gamma,$$

which are to be solved for α, Θ and θ. The solution is

$$\alpha = (1 + \tfrac{1}{4}\gamma^2)^{1/2} + \tfrac{1}{2}\gamma = \cot\psi, \quad \Theta = \tfrac{1}{2}\pi - \psi, \quad \theta = \psi,$$

where

$$\psi = \tfrac{1}{2}\tan^{-1}(2/\gamma) \quad (0 < \psi < \tfrac{1}{4}\pi).$$

Thus the principal stretches are $\lambda_1 = \cot\psi$, $\lambda_2 = \tan\psi$, $\lambda_3 = 1$, and unit vectors specifying the referential and current stretch axes are

$$\boldsymbol{p}_1 = \sin\psi\,\boldsymbol{l} + \cos\psi\,\boldsymbol{m}, \quad \boldsymbol{p}_2 = -\cos\psi\,\boldsymbol{l} + \sin\psi\,\boldsymbol{m}, \quad \boldsymbol{p}_3 = \boldsymbol{n},$$
$$\boldsymbol{q}_1 = \cos\psi\,\boldsymbol{l} + \sin\psi\,\boldsymbol{m}, \quad \boldsymbol{q}_2 = -\sin\psi\,\boldsymbol{l} + \cos\psi\,\boldsymbol{m}, \quad \boldsymbol{q}_3 = \boldsymbol{n}.$$

The orientations of these axes are illustrated in Figure 6. The rotation tensor is given in terms of the vectors $\boldsymbol{p}_i$ and $\boldsymbol{q}_i$ by equation (1.83):

$$\boldsymbol{R} = \boldsymbol{q}_r \otimes \boldsymbol{p}_r = \sin 2\psi(\boldsymbol{l} \otimes \boldsymbol{l} + \boldsymbol{m} \otimes \boldsymbol{m}) + \boldsymbol{n} \otimes \boldsymbol{n} + \cos 2\psi(\boldsymbol{l} \otimes \boldsymbol{m} - \boldsymbol{m} \otimes \boldsymbol{l}).$$

Comparison of this result with (1.79) shows that $\boldsymbol{R}$ represents a rotation of amount $-(\frac{1}{2}\pi - 2\psi)$ about the axis defined by $\boldsymbol{n}$, a conclusion also apparent from Figure 6.

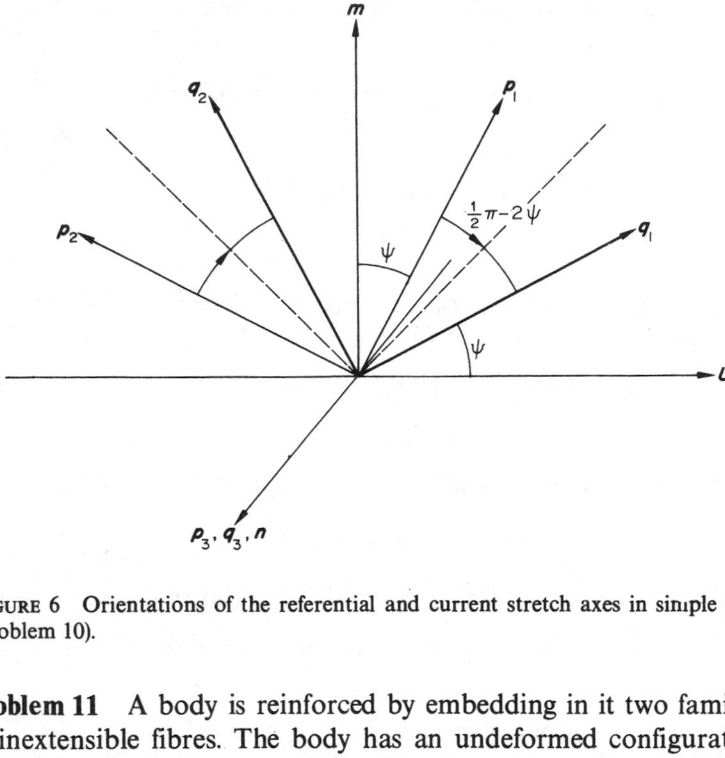

FIGURE 6 Orientations of the referential and current stretch axes in simple shear (Problem 10).

Problem 11 A body is reinforced by embedding in it two families of inextensible fibres. The body has an undeformed configuration B_r in which the fibres in each family are straight and parallel, and unit vectors defining their orientations have components $(\cos \Theta, \pm \sin \Theta, 0)$ $(0 < \Theta < \frac{1}{2}\pi)$ relative to an orthonormal basis E. The body is subjected to a homogeneous triaxial extension in which stretches of amounts $\lambda^{-1/2}\alpha$, $\lambda^{-1/2}\alpha^{-1}$, λ are applied in the directions defined by the base vectors $\boldsymbol{E}_1$, $\boldsymbol{E}_2$, $\boldsymbol{E}_3$ respectively. Obtain an equation connecting α, λ and Θ and deduce from it that

(*a*) the extent to which the body can contract in the 3-direction is limited by the inequality $\lambda \geqslant \sin 2\Theta$;

(*b*) when $\sin 2\Theta < \lambda \neq 1$ two deformed configurations are possible;
(*c*) when the maximum contraction in the 3-direction is achieved, the two families of fibres are orthogonal in the deformed configuration.

Solution. The deformation gradient representing the given triaxial strain is

$$\boldsymbol{F} = \lambda^{-1/2}\alpha \boldsymbol{E}_1 \otimes \boldsymbol{E}_1 + \lambda^{-1/2}\alpha^{-1}\boldsymbol{E}_2 \otimes \boldsymbol{E}_2 + \lambda \boldsymbol{E}_3 \otimes \boldsymbol{E}_3, \quad \text{(A)}$$

and if $\boldsymbol{L}_1$ and $\boldsymbol{L}_2$ are the unit vectors specifying the fibre directions in B_r,

$$\boldsymbol{L}_1 = \cos\Theta\, \boldsymbol{E}_1 + \sin\Theta\, \boldsymbol{E}_2, \quad \boldsymbol{L}_2 = \cos\Theta\, \boldsymbol{E}_1 - \sin\Theta\, \boldsymbol{E}_2. \quad \text{(B)}$$

Since no extension can take place in these directions it follows from equation $(25)_1$ that

$$\boldsymbol{L}_1 \cdot (\boldsymbol{C}\boldsymbol{L}_1) = 1, \quad \boldsymbol{L}_2 \cdot (\boldsymbol{C}\boldsymbol{L}_2) = 1 \quad (\text{with } \boldsymbol{C} = \boldsymbol{F}^{\mathrm{T}}\boldsymbol{F}). \quad \text{(C)}$$

When the expressions (A) and (B) are substituted into (C) we find that in each case

$$\alpha^2 \cos^2\Theta + \alpha^{-2}\sin^2\Theta = \lambda. \quad \text{(D)}$$

This is the required connection between α, λ and Θ. When λ and Θ are given, (D) provides a quadratic equation for α^2 with roots

$$\alpha^2 = \tfrac{1}{2}\sec^2\Theta\{\lambda \pm (\lambda^2 - \sin^2 2\Theta)^{1/2}\}. \quad \text{(E)}$$

(*a*) The roots (E) yield values of α which are real and positive only if $\lambda \geqslant \sin 2\Theta$. This inequality sets a lower bound of $\sin 2\Theta$ to the contraction which can occur in the 3-direction. [Note that contraction is entirely precluded if $\Theta = \frac{1}{4}\pi$.]
(*b*) When $\lambda > \sin 2\Theta$ the roots (E) supply two distinct positive values of α, suggesting that two deformed configurations are possible. However, when $\lambda = 1$, one value of α is unity and no deformation then takes place. Two deformed configurations are possible, therefore, when $\sin 2\Theta < \lambda \neq 1$.
(*c*) From equations (20), with (A) and (B), unit vectors $\boldsymbol{l}_1$ and $\boldsymbol{l}_2$ representing the fibre directions in the deformed configuration are given by

$$\boldsymbol{l}_1 = \boldsymbol{F}\boldsymbol{L}_1 = \lambda^{-1/2}(\alpha\cos\Theta\, \boldsymbol{E}_1 + \alpha^{-1}\sin\Theta\, \boldsymbol{E}_2),$$
$$\boldsymbol{l}_2 = \boldsymbol{F}\boldsymbol{L}_2 = \lambda^{-1/2}(\alpha\cos\Theta\, \boldsymbol{E}_1 - \alpha^{-1}\sin\Theta\, \boldsymbol{E}_2).$$

Each family is seen to be inclined to the 1-direction at an angle θ given by $\tan\theta = \alpha^{-2}\tan\Theta$ or, setting $\alpha^2 = \tan\Theta\cot\theta$ in (D), $\sin 2\theta = \lambda^{-1}\sin 2\Theta$. When the maximum contraction in the 3-direction is attained, $\lambda = \sin 2\Theta$ and $2\theta = \frac{1}{2}\pi$.

[When $\lambda > \sin 2\Theta$ there are two values of θ in the interval $(0, \frac{1}{2}\pi)$, corresponding to the two deformed configurations which are possible: they are given by $\theta = \varphi, \frac{1}{2}\pi - \varphi$ where $\varphi = \frac{1}{2}\sin^{-1}(\lambda^{-1}\sin 2\Theta)$ $(0 < \varphi < \frac{1}{2}\pi)$. Were the body to be very slowly compressed between lubricated rigid plates orthogonal to $\boldsymbol{E}_3$, the deformation in which $\theta = \varphi$ would be realized since $\varphi \to \Theta$, but $\frac{1}{2}\pi - \varphi \nrightarrow \Theta$, as $\lambda \to 1$. In this situation the inclination θ of the fibres to the 1-direction would increase or decrease as λ decreases from unity according as $0 < \Theta < \frac{1}{4}\pi$ or $\frac{1}{4}\pi < \Theta < \frac{1}{2}\pi$, approaching $\frac{1}{4}\pi$ as $\lambda \to \sin 2\Theta$ in each case.]

5 STRETCHING AND SPIN

Whereas the deformation gradient $\boldsymbol{F}$ specifies the changes of size and shape experienced by material elements of a moving body, it is manifest from equations $(31)_1$, (33) and (34) that the velocity gradient $\boldsymbol{L}$ describes the *rate* at which such changes occur. There is, however, a basic distinction between the two gradients which results in the further analysis of $\boldsymbol{L}$ proceeding along lines rather different from those followed in Section 4. This is the fact that $\boldsymbol{F}$ depends upon both the current configuration of a body and a reference configuration while $\boldsymbol{L}$, as the *spatial* gradient of the velocity, is defined without recourse to a reference configuration.

Our present starting point is the unique *additive* decomposition of $\boldsymbol{L}$ into its symmetric and skew-symmetric parts:

$$\boldsymbol{L} = \boldsymbol{D} + \boldsymbol{W} \quad \text{where } \boldsymbol{D} = \tfrac{1}{2}(\boldsymbol{L} + \boldsymbol{L}^{\mathrm{T}}),\ \boldsymbol{W} = \tfrac{1}{2}(\boldsymbol{L} - \boldsymbol{L}^{\mathrm{T}}). \qquad (43)$$

On introducing the polar decompositions (41) into equation $(30)_2$ we find that

$$\boldsymbol{L} = \boldsymbol{R}\dot{\boldsymbol{U}}\boldsymbol{U}^{-1}\boldsymbol{R}^{\mathrm{T}} + \dot{\boldsymbol{R}}\boldsymbol{R}^{\mathrm{T}} = \dot{\boldsymbol{V}}\boldsymbol{V}^{-1} + \boldsymbol{V}\dot{\boldsymbol{R}}\boldsymbol{R}^{\mathrm{T}}\boldsymbol{V}^{-1},$$

and if we recall, from Problem 1 (p. 52), that $\dot{\boldsymbol{R}}\boldsymbol{R}^{\mathrm{T}}$ is skew-symmetric on account of the orthogonality of $\boldsymbol{R}$, the definitions $(43)_{2,3}$ yield the relations

$$\left.\begin{aligned} \boldsymbol{D} &= \tfrac{1}{2}\boldsymbol{R}(\dot{\boldsymbol{U}}\boldsymbol{U}^{-1} + \boldsymbol{U}^{-1}\dot{\boldsymbol{U}})\boldsymbol{R}^{\mathrm{T}} \\ &= \tfrac{1}{2}(\dot{\boldsymbol{V}}\boldsymbol{V}^{-1} + \boldsymbol{V}^{-1}\dot{\boldsymbol{V}} + \boldsymbol{V}\dot{\boldsymbol{R}}\boldsymbol{R}^{\mathrm{T}}\boldsymbol{V}^{-1} - \boldsymbol{V}^{-1}\dot{\boldsymbol{R}}\boldsymbol{R}^{\mathrm{T}}\boldsymbol{V}), \\ \boldsymbol{W} &= \tfrac{1}{2}\boldsymbol{R}(\dot{\boldsymbol{U}}\boldsymbol{U}^{-1} - \boldsymbol{U}^{-1}\dot{\boldsymbol{U}})\boldsymbol{R}^{\mathrm{T}} + \dot{\boldsymbol{R}}\boldsymbol{R}^{\mathrm{T}} \\ &= \tfrac{1}{2}(\dot{\boldsymbol{V}}\boldsymbol{V}^{-1} - \boldsymbol{V}^{-1}\dot{\boldsymbol{V}} + \boldsymbol{V}\dot{\boldsymbol{R}}\boldsymbol{R}^{\mathrm{T}}\boldsymbol{V}^{-1} + \boldsymbol{V}^{-1}\dot{\boldsymbol{R}}\boldsymbol{R}^{\mathrm{T}}\boldsymbol{V}). \end{aligned}\right\} \quad (44)$$

Suppose now that the reference configuration implicit in the definitions of $\boldsymbol{F}$, $\boldsymbol{U}$, $\boldsymbol{V}$ and $\boldsymbol{R}$ is chosen to be the current configuration B_t. Then $\boldsymbol{F}$ is simply the identity tensor and $\boldsymbol{U} = \boldsymbol{V} = \boldsymbol{R} = \boldsymbol{I}$. Equations (44) therefore reduce to

$$\boldsymbol{D} = \dot{\boldsymbol{U}}_0 = \dot{\boldsymbol{V}}_0, \quad \boldsymbol{W} = \dot{\boldsymbol{R}}_0,$$

the subscript o here signifying identification of B_r with B_t. The symmetric and skew-symmetric parts of the velocity gradient are hence the rates of change of the stretch and the rotation as a body passes through its current configuration: $\boldsymbol{D}$ is consequently called the *stretching* (or *rate-of-strain*) *tensor* and $\boldsymbol{W}$ the *spin tensor*.

The geometrical interpretation of stretching and spin in the locality of a representative point x of B_t is based upon equations (33) and (34), and again we refer to the intersecting pair of material arcs shown in Figure 2. Expressed in terms of the stretching tensor, equation $(33)_1$ reads $\boldsymbol{l}.(\boldsymbol{D}\boldsymbol{l}) = (\mathrm{d}s)^{\cdot}/\mathrm{d}s$, while if C_t and D_t meet orthogonally at x, $\boldsymbol{l}.\boldsymbol{m} = 0$ and equation (34) simplifies to $\boldsymbol{l}.(\boldsymbol{D}\boldsymbol{m}) = -\tfrac{1}{2}\dot{\theta}$. The components D_{ij} of the stretching tensor relative to the orthonormal basis e and evaluated at x can thus be given the following meanings. D_{ii} is the rate of extension per unit length of a material line element which, in the current configuration B_t, is situated at x and momentarily aligned with the direction defined by the base vector $\boldsymbol{e}_i$. D_{ij} $(i \neq j)$ is half the rate of decrease of the angle between a pair of material line elements which, in B_t, intersect at x and are instantaneously in the directions represented by $\boldsymbol{e}_i$ and $\boldsymbol{e}_j$.

Since $\boldsymbol{D}$ is a symmetric tensor it possesses three proper numbers ν_i, called the *principal stretchings*, and an orthonormal set of associated proper vectors which define the *principal axes of stretching*. By virtue of the geometrical interpretation just given, the principal stretchings at x are the rates of extension per unit length experienced by material line elements currently aligned with the principal axes of stretching and, moreover, the rates of change of the angles between these lines are momentarily zero. The triplet of material elements in question is therefore instantaneously performing a rigid rotation about x and we show in conclusion that the angular velocity of this rotation is simply the axial vector $\boldsymbol{w}$ of the spin tensor $\boldsymbol{W}$. For a material line element which, in B_t, is situated at x and directed along the principal axis of stretching specified by the unit proper vector $\boldsymbol{r}_i$ of $\boldsymbol{D}$, equation $(33)_2$ gives

$$\begin{aligned}\dot{\boldsymbol{r}}_i &= (\boldsymbol{D} + \boldsymbol{W})\boldsymbol{r}_i - \{\boldsymbol{r}_i . (\boldsymbol{D}\boldsymbol{r}_i)\}\boldsymbol{r}_i \\ &= \boldsymbol{W}\boldsymbol{r}_i + v_i\boldsymbol{r}_i - \{\boldsymbol{r}_i . (v_i\boldsymbol{r}_i)\}\boldsymbol{r}_i = \boldsymbol{W}\boldsymbol{r}_i = \boldsymbol{w} \wedge \boldsymbol{r}_i,\end{aligned}$$

v_i being the principal stretching associated with $\boldsymbol{r}_i$. At time t the element is therefore rotating about x with angular velocity $\boldsymbol{w}$, as stated.

To summarize: *the relative motion in the locality of the point x consists of a triaxial stretching, represented by* $\boldsymbol{D}$, *superimposed upon a rigid rotation specified by* $\boldsymbol{W}$.

Problem 12 A motion in which the velocity is given, in the spatial description, by

$$\boldsymbol{v} = c(\boldsymbol{p} \otimes \boldsymbol{q})\boldsymbol{x} = c(\boldsymbol{x} . \boldsymbol{q})\boldsymbol{p},$$

where c is a positive constant and $\boldsymbol{p}, \boldsymbol{q}$ are orthogonal unit vectors, is called a *simple shearing motion.* For this motion determine the principal stretchings, the principal axes of stretching and the angular velocity.

Solution. This is a steady motion in which the particle paths and streamlines are straight lines in the direction defined by $\boldsymbol{p}$ and material planes orthogonal to $\boldsymbol{q}$ slip over one another without being distorted. Using Exercise 1.14 (i) (p. 49) we obtain

$$\boldsymbol{L} = \operatorname{grad} \boldsymbol{v} = c\boldsymbol{p} \otimes \boldsymbol{q},$$

whence, from $(43)_{2,3}$,

$$\boldsymbol{D} = \tfrac{1}{2}c(\boldsymbol{p} \otimes \boldsymbol{q} + \boldsymbol{q} \otimes \boldsymbol{p}), \quad \boldsymbol{W} = \tfrac{1}{2}c(\boldsymbol{p} \otimes \boldsymbol{q} - \boldsymbol{q} \otimes \boldsymbol{p}).$$

The stretching tensor can be put into the spectral form

$$\boldsymbol{D} = \tfrac{1}{4}c(\boldsymbol{p} + \boldsymbol{q}) \otimes (\boldsymbol{p} + \boldsymbol{q}) - \tfrac{1}{4}c(\boldsymbol{p} - \boldsymbol{q}) \otimes (\boldsymbol{p} - \boldsymbol{q})$$

from which it is plain (by comparison with (1.59)) that the principal stretchings are $\frac{1}{2}c$, $-\frac{1}{2}c$, 0, and that orthonormal vectors specifying the principal axes of stretching are $2^{-1/2}(\boldsymbol{p}+\boldsymbol{q})$, $2^{-1/2}(\boldsymbol{p}-\boldsymbol{q})$, $\boldsymbol{r}$, where $\boldsymbol{r}$ is orthogonal to both $\boldsymbol{p}$ and $\boldsymbol{q}$. Since the sum of the principal stretchings is zero, $\operatorname{tr} \boldsymbol{D} = \operatorname{tr} \boldsymbol{L} = 0$ which means that the motion is isochoric.

Suppose, for definiteness, that the orthonormal basis $\{\boldsymbol{p}, \boldsymbol{q}, \boldsymbol{r}\}$ is positive in E^+ so that the relations (1.69) hold. Let $\boldsymbol{a}$ be an arbitrary vector. Then

$$\begin{aligned} \boldsymbol{W}\boldsymbol{a} &= \tfrac{1}{2}c\{(\boldsymbol{a}\cdot\boldsymbol{q})\boldsymbol{p} - (\boldsymbol{a}\cdot\boldsymbol{p})\boldsymbol{q}\} \\ &= -\tfrac{1}{2}c\boldsymbol{r}\wedge\{(\boldsymbol{a}\cdot\boldsymbol{p})\boldsymbol{p} + (\boldsymbol{a}\cdot\boldsymbol{q})\boldsymbol{q} + (\boldsymbol{a}\cdot\boldsymbol{r})\boldsymbol{r}\} = -\tfrac{1}{2}c\boldsymbol{r}\wedge\boldsymbol{a}, \end{aligned}$$

and we see that $-\frac{1}{2}c\boldsymbol{r}$, as the axial vector of the spin tensor $\boldsymbol{W}$, is the angular velocity. The contributions of the stretching and the spin to the relative motion near the typical point x are illustrated in Figures 7(*a*) and (*b*) respectively.

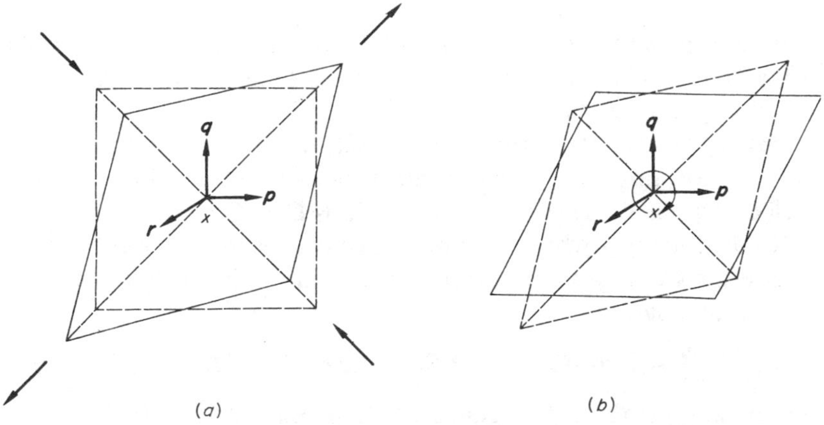

FIGURE 7 Decomposition into stretching (*a*) and spin (*b*) of the deformation over a short time interval of an initially cuboidal material region in simple shearing (Problem 12).

Problem 13 Show that a motion of a body $\mathscr{B}$ is rigid if and only if the stretching tensor is zero in each of the configurations $\{B_t\}$. Deduce that a motion in which no stretch occurs is necessarily rigid.

Solution. (*a*) If $\boldsymbol{D} = \mathbf{O}$ the velocity gradient $\boldsymbol{L}$ coincides with the spin tensor $\boldsymbol{W}$. Furthermore, as we now show, $\boldsymbol{L}$ does not vary with position in B_t.

In relation to a rectangular Cartesian system (o, e) of spatial coordinates the components of $\boldsymbol{L}$ are $\partial v_i/\partial x_j$ (see equations $(29)_1$ and $(1.93)_1$), and the vanishing of $\boldsymbol{D}$ supplies the relation

$$\frac{\partial v_i}{\partial x_j} = -\frac{\partial v_j}{\partial x_i}. \tag{A}$$

Supposing that the velocity components v_i have continuous second partial derivatives, we deduce from (A) that

$$\frac{\partial L_{ij}}{\partial x_k} = \frac{\partial^2 v_i}{\partial x_j \partial x_k} = -\frac{\partial^2 v_j}{\partial x_i \partial x_k} = \frac{\partial^2 v_k}{\partial x_i \partial x_j} = -\frac{\partial^2 v_i}{\partial x_j \partial x_k} = -\frac{\partial L_{ij}}{\partial x_k}.$$

Hence $\partial L_{ij}/\partial x_k = 0$ and $\boldsymbol{L}$ is independent of the spatial position $\boldsymbol{x}$.

We now have that

$$\operatorname{grad} \boldsymbol{v} = \boldsymbol{W}, \tag{B}$$

where $\boldsymbol{W}$ is skew-symmetric and a function of t only. On integrating (B) an expression for $\boldsymbol{v}$ equivalent to $(6)_1$ is obtained and the motion must therefore be rigid.

(*b*) If the motion is rigid the velocity is given by equation $(6)_1$ and on forming the spatial gradient of each side we find that the velocity gradient is skew-symmetric. Thus $\boldsymbol{D} = \mathbf{O}$.

In a motion in which no stretch occurs, the stretch tensors, and hence the Cauchy–Green strain tensors, are universally equal to $\boldsymbol{I}$. From the relation

$$\dot{\boldsymbol{B}} = \dot{\boldsymbol{F}}\boldsymbol{F}^{\mathrm{T}} + \boldsymbol{F}\dot{\boldsymbol{F}}^{\mathrm{T}} = \boldsymbol{L}\boldsymbol{F}\boldsymbol{F}^{\mathrm{T}} + \boldsymbol{F}\boldsymbol{F}^{\mathrm{T}}\boldsymbol{L}^{\mathrm{T}} = \boldsymbol{L}\boldsymbol{B} + \boldsymbol{B}\boldsymbol{L}^{\mathrm{T}},$$

obtained with the aid of equations $(24)_2$ and $(30)_1$, it follows that $\mathbf{O} = \dot{\boldsymbol{B}} = \boldsymbol{L} + \boldsymbol{L}^{\mathrm{T}} = 2\boldsymbol{D}$. In view of the result proved in (*a*) above, the motion is therefore rigid.

6 CIRCULATION AND VORTICITY

The line integral

$$C(\Gamma) = \oint_{\Gamma} \boldsymbol{v} \,.\, \mathrm{d}\boldsymbol{x} \tag{45}$$

of the velocity round a circuit Γ (i.e. a simple closed curve) in the current configuration B_t of a moving body is called the *circulation* round Γ. If Γ is reducible in B_t, that is if there is a regular surface segment Λ having boundary Γ and wholly contained in B_t, we can use Stokes's theorem (equation (1.100)) to rewrite equation (45) as

$$C(\Gamma) = \int_{\Lambda} \boldsymbol{\omega} \cdot \boldsymbol{n} \, da, \quad \text{where } \boldsymbol{\omega} = \operatorname{curl} \boldsymbol{v}, \tag{46}$$

the circuit Γ being positively orientated relative to the unit vector field $\boldsymbol{n}$ normal to Λ. The circulation round Γ is thus equal to the flux of the vector $\boldsymbol{\omega}$ across a surface spanning Γ: $\boldsymbol{\omega}$ is called the *vorticity*. Since $\frac{1}{2}\operatorname{curl} \boldsymbol{v}$ is the axial vector of the spin tensor (see Exercise 1.15, p. 49), the vorticity is equal to twice the local angular velocity.

Formulae for the rate of change of the circulation round a reducible *material* circuit Γ_t follow from equations (38) and (39). Applying (38)[7] and (1.100) successively to the definition (45) we obtain

$$\frac{d}{dt} C(\Gamma_t) = \oint_{\Gamma_t} \boldsymbol{a} \cdot d\boldsymbol{x} = \int_{\Lambda_t} (\operatorname{curl} \boldsymbol{a}) \cdot \boldsymbol{n} \, da, \tag{47}$$

while (39), applied to equation (46), yields

$$\frac{d}{dt} C(\Gamma_t) = \int_{\Lambda_t} (\dot{\boldsymbol{\omega}} + \boldsymbol{\omega} \operatorname{tr} \boldsymbol{L} - \boldsymbol{L}\boldsymbol{\omega}) \cdot \boldsymbol{n} \, da. \tag{48}$$

Here Λ_t is a regular material surface segment with boundary Γ_t. The identity

$$\operatorname{curl} \boldsymbol{a} = \dot{\boldsymbol{\omega}} + \boldsymbol{\omega} \operatorname{tr} \boldsymbol{L} - \boldsymbol{L}\boldsymbol{\omega},$$

evident from equations (47) and (48), can also be derived directly, a task which is left to the reader.

A motion is said to be *circulation preserving* if the circulation round every material circuit is invariable. Motions of this type are of considerable importance in the dynamics of fluids (see Exercise 3.7 (i), p. 122). Bearing in mind the remarks at the end of Problem 1.19 (p. 42), we infer from equations (47) and (48) that a motion is

[7] The integral round Γ of $\boldsymbol{L}^T \boldsymbol{v} = \operatorname{grad}(\frac{1}{2}\boldsymbol{v} \cdot \boldsymbol{v})$ is zero.

circulation preserving if and only if

$$\operatorname{curl} \boldsymbol{a} = \dot{\boldsymbol{\omega}} + \boldsymbol{\omega} \operatorname{tr} \boldsymbol{L} - \boldsymbol{L}\boldsymbol{\omega} = \mathbf{0}. \tag{49}$$

The vorticity equation $(49)_2$ can be recast, with the aid of $(30)_3$ and $(31)_1$, into the form $(J\boldsymbol{F}^{-1}\boldsymbol{\omega})^{\cdot} = \mathbf{0}$ which tells us that the value of $J\boldsymbol{F}^{-1}\boldsymbol{\omega}$ at a representative particle does not change during the motion. In the reference configuration B_r involved in the definition of the deformation gradient, $\boldsymbol{F} = \boldsymbol{I}$ and $J = 1$. Hence

$$\boldsymbol{\omega} = J^{-1}\boldsymbol{F}\boldsymbol{\omega}_r, \tag{50}$$

where $\boldsymbol{\omega}_r$ is the vorticity in B_r. This result, which is known as *Cauchy's vorticity formula*, exhibits a simple relationship between the vorticity fields in two configurations of a circulation preserving motion.

Problem 14 The field lines of the vorticity, given in the spatial description, are called *vortex lines.* Show that, in a circulation preserving motion, the vortex lines are material curves.

Solution. Let C_t be a material curve in B_t with configuration C_r in B_r and suppose that C_r is a vortex line. The problem will be solved if it can be proved that C_t is also a vortex line.

Referring once again to Figure 2, we first recall the relation $\boldsymbol{l}\mathrm{d}s = \boldsymbol{F}\boldsymbol{L}\mathrm{d}S$, $\boldsymbol{L}$, $\mathrm{d}S$ and $\boldsymbol{l}$, $\mathrm{d}s$ being the unit tangent vectors and arc lengths of corresponding elementary arcs of C_r and C_t respectively. Next, since C_r is a field line of the referential vorticity field $\boldsymbol{\omega}_r$, we have $\boldsymbol{\omega}_r = \pm|\boldsymbol{\omega}_r|\boldsymbol{L}$. These results, in combination with the vorticity formula (50), give

$$\boldsymbol{\omega} = \pm J^{-1}|\boldsymbol{\omega}_r|\boldsymbol{F}\boldsymbol{L} = \pm J^{-1}|\boldsymbol{\omega}_r|\,(\mathrm{d}s/\mathrm{d}S)\boldsymbol{l},$$

which shows that the vorticity is directed tangentially to C_t at an arbitrary point. C_t is therefore a vortex line.

[As a rider to this solution we note the relation

$$(\mathrm{d}s/J|\boldsymbol{\omega}|)^{\cdot} = 0, \quad \text{or} \quad (\mathrm{d}s)^{\cdot}/\mathrm{d}s = \operatorname{tr} \boldsymbol{L} + |\boldsymbol{\omega}|^{\cdot}/|\boldsymbol{\omega}|,$$

governing the stretching experienced by a vortex line in a circulation preserving motion.]

If the vorticity $\boldsymbol{\omega}$ (or, equivalently, the spin tensor $\boldsymbol{W}$) is everywhere zero in a region occupied by particles of a moving body, the motion is said to be *irrotational* in that region. It follows from equation (46) that a motion is irrotational in a specified region R if and only if the circulation is zero round every circuit which is reducible *in* R. We speak of an entire motion as being irrotational when $\boldsymbol{\omega}$, or $\boldsymbol{W}$, vanishes in every configuration.

Problem 15 In relation to a rectangular Cartesian coordinate system the velocity field in a certain motion has the spatial description

$$v_1 = -\tfrac{1}{2}\alpha x_1 - f(r)x_2, \quad v_2 = -\tfrac{1}{2}\alpha x_2 + f(r)x_1, \quad v_3 = \alpha x_3, \qquad \text{(A)}$$

where $r = (x_1^2 + x_2^2)^{1/2}$ denotes distance from the 3-axis and α is a positive constant. Describe the nature of the motion and determine the vorticity. Investigate the possibility of the motion being (*a*) irrotational, and (*b*) circulation preserving.

Solution. Equations (A) describe a steady motion obtained by superposing a converging motion in which the velocity components are $(-\tfrac{1}{2}\alpha x_1, -\tfrac{1}{2}\alpha x_2, \alpha x_3)$ and a circulatory motion with velocity components $(-f(r)x_2, f(r)x_1, 0)$. The converging motion brings particles in towards the 3-axis while directing them away from the plane $x_3 = 0$, and this motion possesses a single stagnation point situated at the origin. In the circulatory motion the particles describe with constant speeds $rf(r)$ circles in planes x_3 = constant centred on the 3-axis. It can easily be verified, with reference to equation $(32)_2$, that both the constituent motions are isochoric and the same is true, therefore, of the combined motion. The coincident particle paths[8] and streamlines in the overall motion consist of tapered

[8] A complete description of the motion, obtained by integrating equations (A), is $x_1 = (X_1 \cos\chi - X_2 \sin\chi)\,e^{-\frac{1}{2}\alpha t}$, $x_2 = (X_1 \sin\chi + X_2 \cos\chi)\,e^{-\frac{1}{2}\alpha t}$, $x_3 = X_3\, e^{\alpha t}$, where

$$\chi = \frac{2}{\alpha}\int_{Re^{-\frac{1}{2}\alpha t}}^{R} f(s)\frac{ds}{s} \quad \text{and} \quad R = (X_1^2 + X_2^2)^{\frac{1}{2}}.$$

Here the referential and spatial coordinates belong to a common system and the reference configuration is occupied at $t = 0$. The equation

$$(x_1^2 + x_2^2)x_3 = (X_1^2 + X_2^2)X_3,$$

found by eliminating t, represents a family of material surfaces on which the particle paths are situated (see Figure 8).

helices about the 3-axis (illustrated in Figure 8) and plane spirals (in $x_3 = 0$) converging towards the origin.

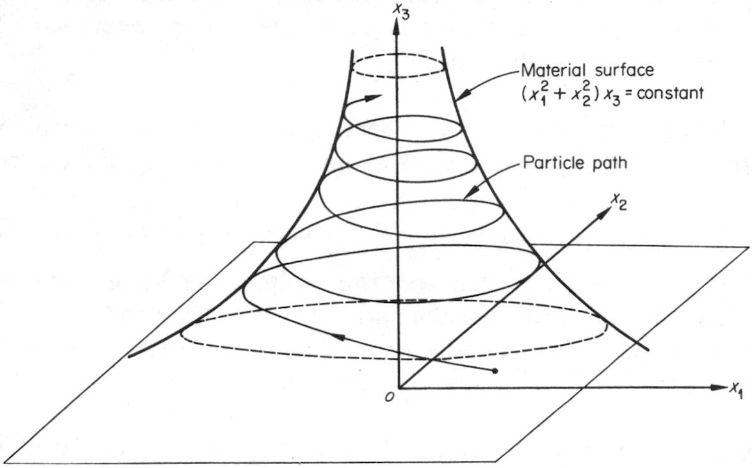

FIGURE 8 A typical particle path and the material surface on which it lies (Problem 15).

The components of vorticity, calculated from (A) with the aid of equations $(46)_2$ and $(1.93)_3$, are

$$\left.\begin{aligned} \omega_1 &= \frac{\partial v_3}{\partial x_2} - \frac{\partial v_2}{\partial x_3} = 0, \quad \omega_2 = \frac{\partial v_1}{\partial x_3} - \frac{\partial v_3}{\partial x_1} = 0, \\ \omega_3 &= \frac{\partial v_2}{\partial x_1} - \frac{\partial v_1}{\partial x_2} = rf'(r) + 2f(r), \end{aligned}\right\} \tag{B}$$

the basis of the given coordinate system being taken to be positive in E^+. The vorticity thus arises entirely from the circulatory motion and is everywhere aligned with the 3-axis.

(*a*) The extent to which the motion is irrotational devolves upon the vanishing of $\boldsymbol{\omega}$ and we see from (B) that f must then satisfy the first-order differential equation $rf'(r) + 2f(r) = 0$. The general solution is

$$f(r) = kr^{-2}, \tag{C}$$

where k is a constant. When f has the form (C) the velocity field is unbounded as $r \to 0$, wherefore the motion contains a singularity known as a *line vortex*. In any region which excludes the vortex the motion is at all times irrotational. However, the circulation round the circle C given by $x_1^2 + x_2^2 = r^2$, x_3 = constant is $2\pi k$, so the motion as a whole is not irrotational.

(*b*) If either of the conditions (49) holds, the motion is circulation preserving. The vanishing of tr$\boldsymbol{L}$, ω_1 and ω_2 and the non-dependence upon x_3 of v_1, v_2 and ω_3 reduce $(49)_2$ to

$$0 = \dot{\omega}_3 - L_{33}\omega_3 = v_1 \frac{\partial \omega_3}{\partial x_1} + v_2 \frac{\partial \omega_3}{\partial x_2} - \frac{\partial v_3}{\partial x_3}\omega_3,$$

and on substituting for v_1, v_2, v_3 and ω_3 from (A) and (B) and removing the non-zero factor $-\frac{1}{2}\alpha$ we arrive at a second-order differential equation for f, namely

$$r^2 f''(r) + 5rf'(r) + 4f(r) = 0.$$

The general solution is

$$f(r) = kr^{-2} + lr^{-2}\ln r, \tag{D}$$

where k and l are constants. As in case (*a*), the velocity field is singular at $r = 0$ and in consequence the smoothness conditions presupposed in the derivation of equation $(49)_2$ are not satisfied. For this reason the conclusion that (D) implies constancy of circulation in respect of material circuits threaded by the 3-axis is not immediately justified. For the material circuit whose current configuration is the circle C specified above, the circulation is

$$2\pi(k + l\ln r) = 2\pi\{k + l(\ln R - \tfrac{1}{2}\alpha t)\},$$

where R is the radius of the circuit at $t = 0$. Thus the motion is circulation preserving only if f has the form (C). As before the vorticity is zero except on the 3-axis where its norm is infinite, the integral of ω_3 over the disc bounded by C being $2\pi k$.

EXERCISES

(The starred exercise contains results which are used in Chapter 3.)

1. A motion of a body is given, for $t \geqslant 0$, by

$$x_1 = X_1 + ktX_3, \quad x_2 = X_2 + ktX_3, \quad x_3 = X_3 - kt(X_1 + X_2),$$

the referential and spatial coordinates relating to a common rectangular Cartesian system and k being a positive constant. Show that the path of an arbitrary particle $\mathscr{X}$ is a straight line orthogonal to oX where o is the origin and X the point occupied by $\mathscr{X}$ in the reference configuration (i.e. at $t = 0$). If B_r is a slab, bounded by the planes $X_1 = \pm A$, verify that the current configuration is also a slab and that the inclination of its faces to the 1-direction tends to $\frac{1}{4}\pi$ as $t \to \infty$.

2. In a rigid motion of a body $\mathscr{B}$ the spatial positions relative to a fixed origin o of the points occupied by four particles $\mathscr{P}$ and $\mathscr{Q}_i$ $(i = 1, 2, 3)$ are respectively $\boldsymbol{c}$ and $\boldsymbol{c} + \boldsymbol{a}_i$, the $\boldsymbol{a}_i$ being orthonormal vectors. Show that the spin tensor $\boldsymbol{W}$ and the angular velocity $\boldsymbol{w}$ of $\mathscr{B}$ are given by

$$\boldsymbol{W} = \tfrac{1}{2}(\dot{\boldsymbol{a}}_p \otimes \boldsymbol{a}_p - \boldsymbol{a}_p \otimes \dot{\boldsymbol{a}}_p), \quad \boldsymbol{w} = \tfrac{1}{2}\boldsymbol{a}_p \wedge \dot{\boldsymbol{a}}_p.$$

If $\mathscr{R}$ is a fifth particle whose place in the current configuration of $\mathscr{B}$ has position $\boldsymbol{c} + \boldsymbol{r}$ relative to o, calculate $\dot{\boldsymbol{r}}$ and $\ddot{\boldsymbol{r}}$.

3. The velocity field in a motion of a body is given, in the spatial description and relative to a rectangular Cartesian coordinate system, by

$$v_1 = -n \sin nt\,(2 + \cos nt)^{-1} x_1,$$
$$v_2 = n \cos nt\,(2 + \sin nt)^{-1} x_2, \quad v_3 = 0,$$

where n is a positive constant. Choosing as reference configuration the placement of the body at $t = 0$ and adopting a common rectangular Cartesian coordinate system, obtain expressions for the x_i in terms of the referential coordinates X_α and t. Determine the particle paths and the streamlines of the motion and discuss the relationship between them. Find the volume v at time t of a material region which has volume V in the reference configuration, and show that the greatest and least values of v are $(\frac{3}{4} \pm \frac{1}{3}\sqrt{2})V$.

4. Find a velocity field for which the ellipsoid given, with reference to a rectangular Cartesian system of spatial coordinates, by

$$\frac{x_1^2}{a_1^2}\left(\frac{t}{\tau}\right)^{\alpha_1} + \frac{x_2^2}{a_2^2}\left(\frac{t}{\tau}\right)^{\alpha_2} + \frac{x_3^2}{a_3^2}\left(\frac{t}{\tau}\right)^{\alpha_3} = 1$$

is a material surface. (Here a_1, a_2, a_3, α_1, α_2, α_3 and τ are positive constants.) Under what condition on the exponents α_1, α_2, α_3 is your motion isochoric?

5. Obtain the formulae

$$(\mathrm{d}a)^{\cdot} = \{\operatorname{tr} \boldsymbol{L} - \boldsymbol{n}\,.\,(\boldsymbol{L}\boldsymbol{n})\}\,\mathrm{d}a, \quad \dot{\boldsymbol{n}} = \{\boldsymbol{n}\,.\,(\boldsymbol{L}\,\boldsymbol{n})\}\boldsymbol{n} - \boldsymbol{L}^T\boldsymbol{n},$$

relating to a material surface element of area da with unit normal $\boldsymbol{n}$, $\boldsymbol{L}$ being the velocity gradient at the current location of the element.

6*. Show that the formulae (37) and (40) can be put into the alternative forms

$$\frac{\mathrm{d}}{\mathrm{d}t}\int_{R_t} \phi\,\mathrm{d}v = \int_{R_t} \frac{\partial \phi}{\partial t}\,\mathrm{d}v + \int_{\partial R_t} \phi \boldsymbol{v}\,.\,\boldsymbol{n}\,\mathrm{d}a,$$

$$\frac{\mathrm{d}}{\mathrm{d}t}\int_{R_t} \boldsymbol{u}\,\mathrm{d}v = \int_{R_t} \frac{\partial \boldsymbol{u}}{\partial t}\,\mathrm{d}v + \int_{\partial R_t} \boldsymbol{u}(\boldsymbol{v}\,.\,\boldsymbol{n})\,\mathrm{d}a,$$

provided that the material region R_t is regular. Here $\boldsymbol{n}$ is the outward unit vector normal to the boundary ∂R_t of R_t.

7. (i) For the simple shear given by equation (42) find the directions in the deformed configuration in which no extension takes place. Find also the directions orthogonal to planes in which no change of area occurs.

(ii) A certain non-rigid isochoric homogeneous deformation is such that any direction orthogonal to planes in which no change of area occurs is also a direction in which no extension takes place. Show that there is at most one such direction and that when it exists it is aligned with a stretch axis.

8. To what extent are the conclusions (*a*), (*b*) and (*c*) of Problem 11 (p. 72) affected by the removal of one of the families of fibres?

9. In a purely circulatory motion the velocity field is given, in the spatial description and relative to a rectangular Cartesian coordinate system, by

$$v_1 = -f(r, t)x_2, \quad v_2 = f(r, t)x_1, \quad v_3 = 0,$$

where $r = (x_1^2 + x_2^2)^{1/2}$. Find the principal stretchings, the principal axes of stretching and the angular velocity at the point with coordinates (x_1, x_2, x_3). Show that this motion is circulation preserving if and only if it is steady.

10. A *steady homogeneous motion* is one for which the velocity is given, in the spatial description, by $\boldsymbol{v} = \boldsymbol{A}\boldsymbol{x}$ where $\boldsymbol{A}$ is a constant tensor. If such a motion is circulation preserving, show that the stretching tensor and the vorticity satisfy the relation $\boldsymbol{D}\boldsymbol{\omega} = (\mathrm{tr}\boldsymbol{D})\boldsymbol{\omega}$. Deduce that two of the principal stretchings have equal magnitudes and opposite signs and that the vorticity is orthogonal to each of the associated principal axes of stretching. Show also that $\boldsymbol{A}^2$ is symmetric.

Chapter 3

BALANCE LAWS, FIELD EQUATIONS AND JUMP CONDITIONS

Classical continuum mechanics rests upon equations expressing the balances of mass, linear momentum, angular momentum and energy in a moving body. These balance laws apply to all material bodies, whether fluid or solid in composition, and each one gives rise to a field equation, holding on the configurations of a body in a sufficiently smooth motion, and a jump condition applying on surfaces of discontinuity. Like position, time and body, the concepts of mass, force, heating and internal energy which enter into the formulation of the balance laws are regarded as having a primitive status in the mechanics of continua to the extent that they are not defined in terms of ideas considered more fundamental.[1] In this chapter we are chiefly concerned with the balance laws and the associated field equations and jump conditions, but prominence is also given to the theory of stress which, besides providing results essential to the development of the field equations, is of basic importance in the analysis of force.

1 MASS

At the intuitive level mass is perceived to be a measure of the amount of material contained in an arbitrary portion of a body. As such it is a non-negative scalar quantity independent of the time and not generally determined by the size of the configuration occupied by the arbitrary sub-body. Furthermore, the mass of a body is the sum of the masses of its parts. These statements can be formalized

[1] But this is not to say that the study of processes contributing, at the atomic level, to the manifestation of force, internal energy, etc., in particular types of materials is not of prime importance.

mathematically by characterizing mass as a set function with certain properties and we proceed on the basis of the following definition.

Let B be an arbitrary configuration of a body $\mathscr{B}$ and let A be a set of points in B occupied by the particles in an arbitrary subset $\mathscr{A}$ of $\mathscr{B}$. If with $\mathscr{A}$ there is associated a non-negative real number $m(\mathscr{A})$ having a physical dimension independent of time and distance and such that

(i) $m(\mathscr{A}_1 \cup \mathscr{A}_2) = m(\mathscr{A}_1) + m(\mathscr{A}_2)$ for all pairs $\mathscr{A}_1$, $\mathscr{A}_2$ of disjoint subsets of $\mathscr{B}$, and

(ii) $m(\mathscr{A}) \to 0$ as the volume of A tends to zero,

$\mathscr{B}$ is said to be a *material body* with *mass function m*. The *mass content* of A, denoted by $m(A)$, is identified with the mass $m(\mathscr{A})$ of $\mathscr{A}$, and we refer henceforth to $\mathscr{A}$ as the material occupying A. Properties (i) and (ii) imply the existence of a scalar field ρ, defined on B, such that

$$m(\mathscr{A}) = m(A) = \int_A \rho \, dv. \tag{1}$$

ρ is called the *mass density*, or simply the *density*, of the material composing $\mathscr{B}$.

In the case of a material body executing a motion $\{B_t \colon t \in I\}$, the density ρ is defined as a scalar field on the configurations $\{B_t\}$ and the mass content $m(R_t)$ of an arbitrary material region R_t in the current configuration B_t is equal to the mass $m(\mathscr{R})$ of the material $\mathscr{R}$ occupying R_t at time t. Since $m(\mathscr{R})$ does not depend upon t we deduce directly from (1) the *equation of mass balance*

$$\frac{d}{dt}\int_{R_t} \rho \, dv = 0. \tag{2}$$

The density is assumed to be continuously differentiable jointly in the position and time variables on which it depends. The transport formula (2.37) can therefore be used to transform equation (2) into

$$\int_{R_t} (\dot{\rho} + \rho \operatorname{div} \boldsymbol{v}) \, dv = 0,$$

wherein the integrand is continuous in B_t and the range of integration is an arbitrary subregion of B_t. Thus, by Problem 1.19 (p. 42), the integrand vanishes in B_t and we arrive at the *spatial equation of continuity*,

$$\dot{\rho} + \rho \operatorname{div} \boldsymbol{v} = 0. \tag{3}$$

Subject to the presumed smoothness of ρ, equation (3) is a consequence of (2) and the reverse implication obviously holds. The *balance law* (2) and the *field equation* (3) are therefore equivalent expressions of the principle of mass conservation.

Problem 1 Let ϕ be a scalar field and $\boldsymbol{u}$ a vector field representing properties of a moving material body $\mathscr{B}$, and let R_t be an arbitrary material region in the current configuration of $\mathscr{B}$. Show that

$$\frac{\mathrm{d}}{\mathrm{d}t}\int_{R_t} \rho\phi \, \mathrm{d}v = \int_{R_t} \rho\dot{\phi} \, \mathrm{d}v, \tag{4}$$

$$\frac{\mathrm{d}}{\mathrm{d}t}\int_{R_t} \rho\boldsymbol{u} \, \mathrm{d}v = \int_{R_t} \rho\dot{\boldsymbol{u}} \, \mathrm{d}v. \tag{5}$$

Solution. The result of replacing the scalar field ϕ by $\rho\phi$ in equation (2.37) is

$$\begin{aligned}\frac{\mathrm{d}}{\mathrm{d}t}\int_{R_t} \rho\phi \, \mathrm{d}v &= \int_{R_t} \{(\rho\phi)^{\cdot} + \rho\phi \operatorname{tr} \boldsymbol{L}\} \, \mathrm{d}v \\ &= \int_{R_t} \{\rho\dot{\phi} + (\dot{\rho} + \rho \operatorname{div} \boldsymbol{v})\phi\} \, \mathrm{d}v,\end{aligned}$$

and on using equation (3) we obtain the first of the required results. Equation (2.40) similarly yields the vector formula (5).

Problem 2 Establish the relation

$$\rho\boldsymbol{a} = \frac{\partial}{\partial t}(\rho\boldsymbol{v}) + \operatorname{div}(\rho\boldsymbol{v} \otimes \boldsymbol{v})$$

connecting the density, the velocity and the acceleration of a moving material body.

Solution. Let R_t be a regular material region in the current configuration of the body. Then, writing $\rho\boldsymbol{v}$ in place of $\boldsymbol{u}$ in the second result of Exercise 2.6 (p. 85), we have

$$\frac{\mathrm{d}}{\mathrm{d}t}\int_{R_t} \rho\boldsymbol{v} \, \mathrm{d}v - \int_{R_t} \frac{\partial}{\partial t}(\rho\boldsymbol{v}) \, \mathrm{d}v = \int_{\partial R_t} \rho(\boldsymbol{n} \cdot \boldsymbol{v})\boldsymbol{v} \, \mathrm{d}a = \int_{\partial R_t} \rho(\boldsymbol{v} \otimes \boldsymbol{v})\boldsymbol{n} \, \mathrm{d}a.$$

When equation (5) is applied to the first term on the left and the divergence theorem, in the form (1.96), to the surface integral on the right there follows

$$\int_{R_t} \left\{ \rho \boldsymbol{a} - \frac{\partial}{\partial t}(\rho \boldsymbol{v}) - \operatorname{div}(\rho \boldsymbol{v} \otimes \boldsymbol{v}) \right\} dv = \boldsymbol{0},$$

and the solution is completed by utilizing Problem 1.19 (p. 42), as in the derivation of equation (3).

The elimination of div $\boldsymbol{v}$ between equations (3) and (2.31) yields $(\rho J)^{\cdot} = 0$, indicating that the value of ρJ at a given particle does not change during the motion of a material body. Let ρ_r be the density in the reference configuration B_r involved in the definition of the deformation gradient. Then $\rho = \rho_r$ and $J = 1$ in B_r, giving ρ_r as the constant value of ρJ and hence the *referential equation of continuity*

$$\rho = J^{-1}\rho_r. \tag{6}$$

This result is also a direct consequence of equation (2.22), the material volume element shown in Figure 2.4 having mass content $\rho_r dV$ in B_r and ρdv in B_t.

A material body which is able to undergo only isochoric motions is said to be composed of *incompressible material.* Since $J = 1$ in an isochoric motion, we see from equation (6) that, at an arbitrary particle of an incompressible material, the density is invariable. If, in particular, the density is uniform in some configuration it has the same uniform value in every configuration which the body can occupy.

2 MOMENTUM, FORCE AND TORQUE

At this point we turn away from the purely kinematic and inertial aspects of material bodies to the causes of motion and the principles governing them.

Let R_t be an arbitrary material region in the configuration B_t assumed at time t by a moving material body $\mathscr{B}$. The *linear momentum, $\boldsymbol{M}(R_t)$,* of the material occupying R_t is defined by

$$\boldsymbol{M}(R_t) = \int_{R_t} \rho \boldsymbol{v} \, dv. \tag{7}$$

If x is the position of a representative point of R_t relative to an origin o the *angular momentum with respect to o*, $H(R_t; o)$, of the same material is defined by

$$H(R_t; o) = \int_{R_t} x \wedge (\rho v)\, dv. \tag{8}$$

Problem 3 Calculate $M(R_t)$ and $H(R_t; o)$ for a material body performing a rigid motion.

Solution. The velocity in a rigid motion is given, in the spatial description, by equation $(2.7)_1$. Thus, from the definitions (7) and (8),

$$M(R_t) = \int_{R_t} \rho\{\dot{c} + w \wedge (x - c)\}\, dv = m(R_t)\{\dot{c} + w \wedge (\bar{x} - c)\} \tag{A}$$

and

$$\begin{aligned} H(R_t; o) &= \int_{R_t} \rho x \wedge \{\dot{c} + w \wedge (x - c)\}\, dv \\ &= m(R_t)\bar{x} \wedge (\dot{c} - w \wedge c) + J(R_t)w, \end{aligned} \tag{B}$$

Here

$$\bar{x} = \{m(R_t)\}^{-1} \int_{R_t} \rho x\, dv, \quad J(R_t) = \int_{R_t} \rho\{(x \cdot x)I - x \otimes x\}\, dv,$$

and use has been made of equations (1), $(1.73)_2$ and (1.41). $\bar{x}$ and $J(R_t)$ are respectively the position of the mass centre and the inertia tensor relative to o of the material occupying R_t.

The results (A) and (B) become more informative when rewritten as

$$M(R_t) = m(R_t)\bar{v}, \quad H(R_t; o) = m(R_t)\bar{x} \wedge \bar{v} + \bar{J}(R_t)w,$$

$\bar{v} = \dot{c} + w \wedge (\bar{x} - c)$ being the velocity of the mass centre and $\bar{J}(R_t)$ the inertia tensor relative to the mass centre.

The concepts of force and torque describe the action on a moving material body $\mathscr{B}$ of its surroundings and the mutual actions of the parts of $\mathscr{B}$ on one another. The mathematical embodiments of these concepts are vector-valued functions defined on subsets of $\mathscr{B}$. Thus, in the configuration B_t, there are associated with an arbitrary material region R_t vectors $\boldsymbol{F}(R_t)$ and $\boldsymbol{G}(R_t; o)$, referred to respectively as the *force* and the *torque with respect to o* acting on the material occupying R_t. Two distinct types of forces and torques are recognized in continuum mechanics: *body forces and torques* are conceived as acting on the particles of a body, and *contact forces and torques* as arising from the action of one part of a body on an adjacent part across a separating surface.

The body force and torque acting on $\mathscr{B}$ are specified by vector fields $\boldsymbol{b}$ and $\boldsymbol{c}$ defined on the configurations of $\mathscr{B}$. These fields are taken as measured per unit mass *and are assumed to be continuous.* Their contributions to $\boldsymbol{F}(R_t)$ and $\boldsymbol{G}(R_t; o)$ are therefore in turn

$$\int_{R_t} \rho \boldsymbol{b} \, dv \quad \text{and} \quad \int_{R_t} \{\boldsymbol{x} \wedge (\rho \boldsymbol{b}) + \rho \boldsymbol{c}\} \, dv. \tag{9}$$

Examples of body forces are the actions on a body arising from its presence in a gravitational or an electromagnetic field.

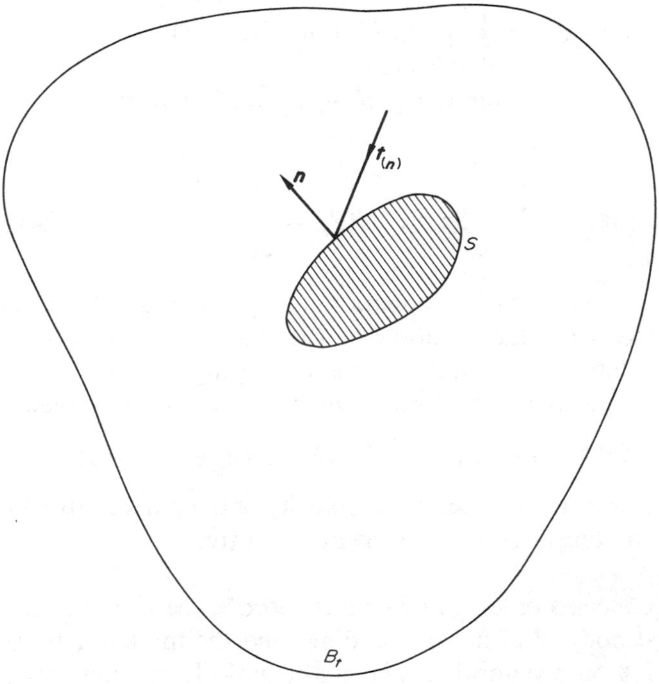

FIGURE 1 The stress vector $\boldsymbol{t}_{(\boldsymbol{n})}$ acting on a surface S with outward unit normal vector $\boldsymbol{n}$.

A mathematical description of contact forces stems from the following *stress principle of Euler and Cauchy*: the action of the material occupying the part of B_t exterior to a closed surface S on the material occupying the interior part is represented by a vector field $\boldsymbol{t}_{(\boldsymbol{n})}$, with physical dimension force per unit area, defined on S (see Figure 1). $\boldsymbol{t}_{(\boldsymbol{n})}$ is called the *stress vector and is assumed to depend continuously upon* $\boldsymbol{n}$, the outward unit vector normal to S.

A material body for which a complete account of contact action is given by the preceding statement is said to be composed of *non-polar material.* Such material admits contact forces but not contact torques. The classical branches of continuum mechanics are exclusively concerned with non-polar materials and we give no direct consideration to polar media in this book.

The contributions to $\boldsymbol{F}(R_t)$ and $\boldsymbol{G}(R_t; o)$ of the contact forces acting on the boundary ∂R_t of the arbitrary material region R_t are

$$\int_{\partial R_t} \boldsymbol{t}_{(\boldsymbol{n})}\, da \quad \text{and} \quad \int_{\partial R_t} \boldsymbol{x} \wedge \boldsymbol{t}_{(\boldsymbol{n})}\, da$$

respectively. Thus, combining these expressions with (9),

$$\left.\begin{aligned} \boldsymbol{F}(R_t) &= \int_{R_t} \rho \boldsymbol{b}\, \mathrm{d}v + \int_{\partial R_t} \boldsymbol{t}_{(\boldsymbol{n})}\, da, \\ \boldsymbol{G}(R_t; o) &= \int_{R_t} \rho(\boldsymbol{x} \wedge \boldsymbol{b} + \boldsymbol{c})\, \mathrm{d}v + \int_{\partial R_t} \boldsymbol{x} \wedge \boldsymbol{t}_{(\boldsymbol{n})}\, da. \end{aligned}\right\} \tag{10}$$

The *laws of motion*, first enunciated by Euler, apply to all bodies composed of non-polar material. They assert the existence of a frame in which, regardless of the choice of the origin o,

$$\frac{\mathrm{d}}{\mathrm{d}t}\boldsymbol{M}(R_t) = \boldsymbol{F}(R_t), \quad \frac{\mathrm{d}}{\mathrm{d}t}\boldsymbol{H}(R_t; o) = \boldsymbol{G}(R_t; o), \tag{11}$$

the term *frame* signifying a system of measuring position in $\mathfrak{E}$ (i.e. a correspondence between $\mathfrak{E}$ and the supporting vector space E of the kind described in Section 1.9) together with a means of measuring time.

Problem 4 Show that

$$\boldsymbol{G}(R_t; y) = \boldsymbol{G}(R_t; o) - \boldsymbol{y} \wedge \boldsymbol{F}(R_t),$$

where y is an arbitrary point and $\boldsymbol{y} = \overrightarrow{oy}$.

Solution. According to the second law of motion,

$$\boldsymbol{G}(R_t; y) = (\mathrm{d}/\mathrm{d}t)\boldsymbol{H}(R_t; y),$$

and from the definitions (8) and (7),

$$\begin{aligned}\boldsymbol{H}(R_t; y) &= \int_{R_t} (\boldsymbol{x} - \boldsymbol{y}) \wedge (\rho\boldsymbol{v})\,\mathrm{d}v = \int_{R_t} \boldsymbol{x} \wedge (\rho\boldsymbol{v})\,\mathrm{d}v - \boldsymbol{y} \wedge \int_{R_t} \rho\boldsymbol{v}\,\mathrm{d}v \\ &= \boldsymbol{H}(R_t; o) - \boldsymbol{y} \wedge \boldsymbol{M}(R_t).\end{aligned}$$

Hence, with the use of equations (11),

$$\begin{aligned}\boldsymbol{G}(R_t; y) &= (\mathrm{d}/\mathrm{d}t)\boldsymbol{H}(R_t; o) - \boldsymbol{y} \wedge \{(\mathrm{d}/\mathrm{d}t)\boldsymbol{M}(R_t)\} \\ &= \boldsymbol{G}(R_t; o) - \boldsymbol{y} \wedge \boldsymbol{F}(R_t).\end{aligned}$$

[In words, the torque with respect to y acting on the material occupying R_t equals the torque with respect to o plus the moment about y of the force on this material, regarded as acting through o. The same proposition can be proved directly from equations (10), but the argument used here involves no assumption about the character of the forces and torques contributing to $\boldsymbol{F}$ and $\boldsymbol{G}$.]

Body torques do not arise in the situations commonly studied by the methods of classical continuum mechanics and they are assumed henceforth to be absent. The integral relations obtained by substituting from equations (7), (8) and (10) into (11) are therefore

$$\left.\begin{aligned}\frac{\mathrm{d}}{\mathrm{d}t}\int_{R_t} \rho\boldsymbol{v}\,\mathrm{d}v &= \int_{R_t} \rho\boldsymbol{b}\,\mathrm{d}v + \int_{\partial R_t} \boldsymbol{t}_{(\boldsymbol{n})}\,\mathrm{d}a, \\ \frac{\mathrm{d}}{\mathrm{d}t}\int_{R_t} \rho\boldsymbol{x} \wedge \boldsymbol{v}\,\mathrm{d}v &= \int_{R_t} \rho\boldsymbol{x} \wedge \boldsymbol{b}\,\mathrm{d}v + \int_{\partial R_t} \boldsymbol{x} \wedge \boldsymbol{t}_{(\boldsymbol{n})}\,\mathrm{d}a,\end{aligned}\right\} \qquad (12)$$

and in this guise the laws of motion are recognized as equations of linear and angular momentum balance. With the application of the vector formula (5) to their left-hand sides equations (12) become

$$\left.\begin{aligned}\int_{R_t} \rho(\boldsymbol{a} - \boldsymbol{b})\,\mathrm{d}v &= \int_{\partial R_t} \boldsymbol{t}_{(\boldsymbol{n})}\,\mathrm{d}a, \\ \int_{R_t} \rho\boldsymbol{x} \wedge (\boldsymbol{a} - \boldsymbol{b})\,\mathrm{d}v &= \int_{\partial R_t} \boldsymbol{x} \wedge \boldsymbol{t}_{(\boldsymbol{n})}\,\mathrm{d}a,\end{aligned}\right\} \qquad (13)$$

and on forming the vector product of each side of $(13)_2$ with an arbitrary vector and using equation $(1.73)_1$ we see that it is permissible to replace the integrands in this equation by the skew-symmetric tensors of which they are the axial vectors. Thus

$$\int_{R_t} \rho\{\boldsymbol{x} \otimes (\boldsymbol{a} - \boldsymbol{b}) - (\boldsymbol{a} - \boldsymbol{b}) \otimes \boldsymbol{x}\}\,dv = \int_{\partial R_t} (\boldsymbol{x} \otimes \boldsymbol{t}_{(\boldsymbol{n})} - \boldsymbol{t}_{(\boldsymbol{n})} \otimes \boldsymbol{x})\,da. \tag{14}$$

3 THE THEORY OF STRESS

As a necessary prelude to the derivation from $(13)_1$ and (14) of field equations expressing the local balances of momenta, a deeper analysis of contact force must next be undertaken. The central conclusion of the resulting theory,[2] reached by applying equation $(13)_1$ to a region in the form of a tetrahedron, is the existence of a *stress tensor* $\boldsymbol{\sigma}$ which is connected to the stress vector $\boldsymbol{t}_{(\boldsymbol{n})}$ and the unit normal vector $\boldsymbol{n}$ by the relation

$$\boldsymbol{t}_{(\boldsymbol{n})} = \boldsymbol{\sigma}^{\mathrm{T}}\boldsymbol{n}. \tag{15}$$

The stress tensor is defined on the configurations of the material body under consideration and does not depend upon $\boldsymbol{n}$. Accordingly, equation (15) makes explicit the dependence of the stress vector on the unit normal.

An element of area da on the surface S shown in Figure 1 is subject to contact forces $\boldsymbol{t}_{(\boldsymbol{n})}\,da$ and $\boldsymbol{t}_{(-\boldsymbol{n})}\,da = -\boldsymbol{t}_{(\boldsymbol{n})}\,da$ on its inner and outer sides as shown in Figure 2. The normal component of the stress vector,

$$\sigma_{(\boldsymbol{n})} = \boldsymbol{t}_{(\boldsymbol{n})}\cdot\boldsymbol{n} = \boldsymbol{n}\cdot(\boldsymbol{\sigma}\boldsymbol{n}), \tag{16}$$

is called the *normal stress* and is said to be *tensile* when positive and *compressive* when negative. The component of the stress vector directed tangentially to S has norm

$$\tau_{(\boldsymbol{n})} = |\boldsymbol{t}_{(\boldsymbol{n})} - (\boldsymbol{t}_{(\boldsymbol{n})}\cdot\boldsymbol{n})\boldsymbol{n}| = (|\boldsymbol{t}_{(\boldsymbol{n})}|^2 - \sigma_{(\boldsymbol{n})}^2)^{1/2} \tag{17}$$

and is known as the *shear stress*. If, in some configuration, the shear

[2] See, for example, W. Jaunzemis, *Continuum Mechanics* (Macmillan, New York, 1967), p. 204 *et seq.*

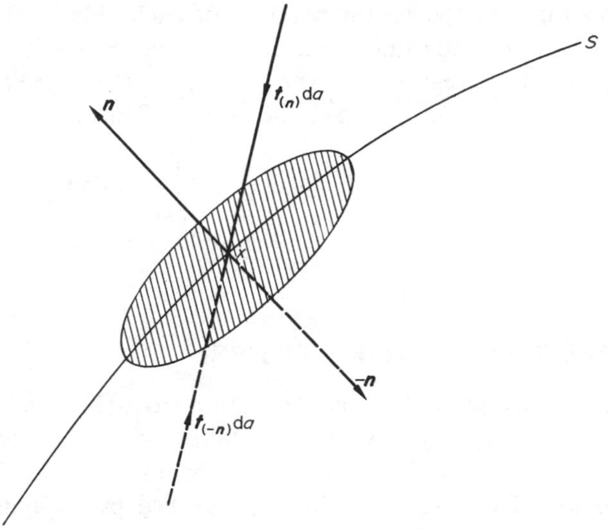

FIGURE 2 Equal and opposite contact forces acting on the two sides of a surface element.

stress is identically zero and the normal stress is independent of $\boldsymbol{n}$, the stress is said to be *spherical*.[3] In this case there is a scalar field p, called the *pressure*, such that

$$\boldsymbol{t}_{(\boldsymbol{n})} = -p\boldsymbol{n} \quad \text{and} \quad \boldsymbol{\sigma} = -p\boldsymbol{I}. \tag{18}$$

At a representative point x in the current configuration B_t let σ_{ij} be the components of the stress tensor relative to an orthonormal basis $e = \{\boldsymbol{e}_1, \boldsymbol{e}_2, \boldsymbol{e}_3\}$. Then, with the use of (15),

$$\sigma_{ij} = \boldsymbol{e}_i \cdot (\boldsymbol{\sigma}\boldsymbol{e}_j) = \boldsymbol{e}_j \cdot (\boldsymbol{\sigma}^{\mathrm{T}}\boldsymbol{e}_i) = \boldsymbol{t}_{(\boldsymbol{e}_i)} \cdot \boldsymbol{e}_j, \tag{19}$$

and σ_{ij} is identified as the j-component of the force per unit area in B_t acting on a surface segment whose outward normal at x is in the i-direction.

Problem 5 With reference to a rectangular Cartesian system (o, e) of spatial coordinates, the configuration B of a material body is given by $-a \leqslant x_1 \leqslant a$, $-a \leqslant x_2 \leqslant a$, $-h \leqslant x_3 \leqslant h$ and the stress field in B

[3] The widely used term *hydrostatic* is avoided here because of its special connotations.

has components

$$\sigma_{11} = -q(x_1^2 - x_2^2)/a^2, \quad \sigma_{22} = q(x_1^2 - x_2^2)/a^2, \quad \sigma_{33} = 0,$$

$$\sigma_{23} = \sigma_{32} = 0, \quad \sigma_{31} = \sigma_{13} = 0, \quad \sigma_{12} = \sigma_{21} = 2qx_1x_2/a^2,$$

a, h and q being positive constants. Determine the tractions (i.e. the forces per unit area) which must be applied to the boundary of B to maintain this state of stress and calculate the resultant force and the resultant torque with respect to o acting on each of the faces $x_1 = \pm a$, $x_2 = \pm a$.

Solution. On the face $x_1 = a$ the outward unit normal coincides with the base vector $\boldsymbol{e}_1$ and the stress vector has components $(\sigma_{11}, \sigma_{12}, \sigma_{13})$. On the opposite face, $x_1 = -a$, the outward unit normal is $-\boldsymbol{e}_1$ and the components of the stress vector are $(-\sigma_{11}, -\sigma_{12}, -\sigma_{13})$. Dealing similarly with the faces $x_2 = \pm a$, $x_3 = \pm h$, we arrive at the six sets of components marked on Figure 3. These

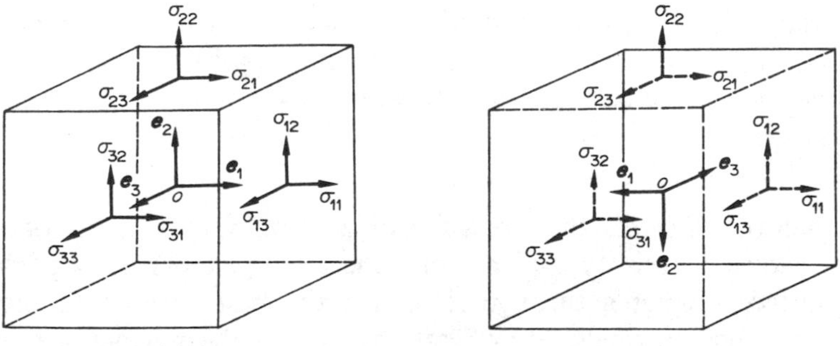

FIGURE 3 Tractions acting on the faces of a cuboid (Problem 5). (*a*) shows the tractions on the faces $x_1 = a$, $x_2 = a$, $x_3 = h$, and (*b*) the tractions on the opposite faces.

components specify the surface tractions which must be supplied in order to support the specified state of stress and they are found by evaluating the given formulae on the appropriate boundaries. Thus

$$\sigma_{11} = -q(a^2 - x_2^2)/a^2, \quad \sigma_{12} = \pm 2qx_2/a, \quad \sigma_{13} = 0 \quad \text{on} \quad x_1 = \pm a,$$

$$\sigma_{21} = \pm 2qx_1/a, \quad \sigma_{22} = q(x_1^2 - a^2)/a^2, \quad \sigma_{23} = 0 \quad \text{on} \quad x_2 = \pm a,$$

and the faces $x_3 = \pm h$ are traction-free.

The resultant force and the resultant torque with respect to o on the face $x_1 = a$ are given by

$$\int_{-a}^{a}\int_{-h}^{h} \sigma_{1p}\Big|_{x_1=a} e_p\, dx_2\, dx_3 = 2h\int_{-a}^{a}\left(-q\frac{a^2-x_2^2}{a^2}e_1 + 2q\frac{x_2}{a}e_2\right)dx_2$$

and

$$\int_{-a}^{a}\int_{-h}^{h} (ae_1 + x_2e_2 + x_3e_3) \wedge \left(\sigma_{1p}\Big|_{x_1=a} e_p\right) dx_2\, dx_3$$

$$= \int_{-a}^{a}\int_{-h}^{h}\left(-2q\frac{x_2x_3}{a}e_1 - q\frac{a^2-x_2^2}{a^2}x_3e_2 + q\frac{3a^2-x_2^2}{a^2}x_2e_3\right)dx_2\, dx_3,$$

the basis e being assumed positive in E^+. On carrying out the integrations the resultants are found to be $-\frac{8}{3}qahe_1$ and $\mathbf{0}$ respectively. Similar calculations show that the resultant forces on the faces $x_1 = -a$, $x_2 = a$ and $x_2 = -a$ are in turn $\frac{8}{3}qahe_1$, $-\frac{8}{3}qahe_2$ and $\frac{8}{3}qahe_2$, while the resultant torque with respect to o is in each case zero.

Problem 6 Express the stress tensor at a point x in terms of an orthonormal basis $\{e_1, e_2, e_3\}$ and the stress vectors on a triplet of surface segments through x having the base vectors as their outward unit normals. Show that the sum of the normal stresses on these segments and the sum of squares of the norms of the stress vectors are both independent of the choice of basis.

Solution. Relative to the given basis the stress tensor has the representation $\sigma = \sigma_{pq}e_p \otimes e_q$ and the components σ_{ij} are given in terms of the base vectors and the stress vectors $t_{(e_i)}$ by equation (19). Hence

$$\sigma = \{t_{(e_p)} \cdot e_q\}\, e_p \otimes e_q = e_p \otimes t_{(e_p)},$$

which is the required expression.

From equation (16) we obtain the sum of the normal stresses at x

on surface segments having outward unit normals $\boldsymbol{e}_1, \boldsymbol{e}_2, \boldsymbol{e}_3$, as

$$\sigma_{(\boldsymbol{e}_1)} + \sigma_{(\boldsymbol{e}_2)} + \sigma_{(\boldsymbol{e}_3)} = \boldsymbol{e}_p \cdot (\boldsymbol{\sigma}\boldsymbol{e}_p) = \sigma_{pp} = \operatorname{tr}\boldsymbol{\sigma},$$

and with the use of (15) the sum of squares of the norms of the stress vectors is found to be

$$\begin{aligned}|\boldsymbol{t}_{(\boldsymbol{e}_1)}|^2 + |\boldsymbol{t}_{(\boldsymbol{e}_2)}|^2 + |\boldsymbol{t}_{(\boldsymbol{e}_3)}|^2 &= \boldsymbol{t}_{(\boldsymbol{e}_p)} \cdot \boldsymbol{t}_{(\boldsymbol{e}_p)} = (\boldsymbol{\sigma}^{\mathrm{T}}\boldsymbol{e}_p) \cdot (\boldsymbol{\sigma}^{\mathrm{T}}\boldsymbol{e}_p) \\ &= \boldsymbol{e}_p \cdot (\boldsymbol{\sigma}\boldsymbol{\sigma}^{\mathrm{T}}\boldsymbol{e}_p) = \operatorname{tr}(\boldsymbol{\sigma}\boldsymbol{\sigma}^{\mathrm{T}}).\end{aligned}$$

In each sum the choice of basis is seen to be immaterial.

When the need arises to distinguish it from other stress tensors $\boldsymbol{\sigma}$ is referred to as the *Cauchy stress.* An alternative measure of stress which has both practical and theoretical significance is the *nominal stress*, $\boldsymbol{s}$, defined by

$$\boldsymbol{s} = J\boldsymbol{F}^{-1}\boldsymbol{\sigma}, \quad \boldsymbol{\sigma} = J^{-1}\boldsymbol{F}\boldsymbol{s}. \tag{20}$$

Whereas the Cauchy stress is concerned exclusively with the current configuration B_t of a material body, the nominal stress, through the appearance of the deformation gradient, involves also a reference configuration B_r. The component form of $\boldsymbol{s}$ relative to rectangular Cartesian systems (O, E) and (o, e) of referential and spatial coordinates is

$$\boldsymbol{s} = s_{\pi p}\boldsymbol{E}_\pi \otimes \boldsymbol{e}_p \quad \text{where} \quad s_{\alpha i} = JF^{-1}_{\alpha p}\sigma_{pi}.$$

An interpretation of the nominal stress is secured by considering a material surface element with area $\mathrm{d}a$ and outward unit normal $\boldsymbol{n}$ in B_t and corresponding properties $\mathrm{d}A$ and $\boldsymbol{N}$ in B_r (see Figure 2.3). The contact force acting on this element in B_t is $\boldsymbol{\sigma}^{\mathrm{T}}\boldsymbol{n}\,\mathrm{d}a$ and, from equation $(20)_2$ in conjunction with the kinematic formula (2.21),

$$\boldsymbol{\sigma}^{\mathrm{T}}\boldsymbol{n}\,\mathrm{d}a = (J^{-1}\boldsymbol{F}\boldsymbol{s})^{\mathrm{T}}\{J(\boldsymbol{F}^{-1})^{\mathrm{T}}\boldsymbol{N}\,\mathrm{d}A\} = \boldsymbol{s}^{\mathrm{T}}\boldsymbol{N}\,\mathrm{d}A. \tag{21}$$

The nominal stress therefore specifies the contact force in B_t per unit referential area and with respect to the orientation in B_r of the surface involved. At the representative point x in B_t, shown in Figure

2.3, and in relation to the coordinate systems (O, E) and (o, e), $s_{\alpha i} = (\boldsymbol{s}^{\mathrm{T}}\boldsymbol{E}_{\alpha}) \cdot \boldsymbol{e}_i$ is the i-component of the contact force per unit area in B_r acting on a material surface segment whose configuration in B_r has its outward normal (at X) in the α-direction.

Two further measures of stress which are sometimes useful are the Piola–Kirchhoff stress tensors. The first of these is $\boldsymbol{s}^T$, the transpose of the nominal stress. The *second Piola–Kirchhoff stress*, $\boldsymbol{t}$, is defined by

$$\boldsymbol{t} = J\boldsymbol{F}^{-1}\boldsymbol{\sigma}(\boldsymbol{F}^{-1})^{\mathrm{T}}, \quad \boldsymbol{\sigma} = J^{-1}\boldsymbol{F}\boldsymbol{t}\boldsymbol{F}^{\mathrm{T}}. \tag{22}$$

Unfortunately $\boldsymbol{t}$ does not admit a direct interpretation, of the kind given above for the Cauchy and nominal stresses, in terms of the contact force acting on a surface segment.

4 EQUATIONS OF MOTION

Returning now to equations $(13)_1$ and (14) and making use of (15) we have

$$\int_{R_t} \rho(\boldsymbol{a} - \boldsymbol{b})\,\mathrm{d}v = \int_{\partial R_t} \boldsymbol{\sigma}^{\mathrm{T}}\boldsymbol{n}\,\mathrm{d}a,$$

$$\int_{R_t} \rho\{\boldsymbol{x} \otimes (\boldsymbol{a} - \boldsymbol{b}) - (\boldsymbol{a} - \boldsymbol{b}) \otimes \boldsymbol{x}\}\,\mathrm{d}v = \int_{\partial R_t} \{\boldsymbol{x} \otimes (\boldsymbol{\sigma}^{\mathrm{T}}\boldsymbol{n}) - (\boldsymbol{\sigma}^{\mathrm{T}}\boldsymbol{n}) \otimes \boldsymbol{x}\} \times \mathrm{d}a.$$

It is assumed that the stress tensor $\boldsymbol{\sigma}$ is continuously differentiable jointly in the position and time variables on which it depends. Taking the material region R_t to be regular we can therefore use the divergence theorem in the forms (1.96) and (1.95) to transform the right-hand sides of these equations into volume integrals, obtaining

$$\int_{R_t} \{\rho(\boldsymbol{a} - \boldsymbol{b}) - \operatorname{div}\boldsymbol{\sigma}\}\,\mathrm{d}v = \boldsymbol{0},$$

$$\int_{R_t} [\boldsymbol{x} \otimes \{\rho(\boldsymbol{a} - \boldsymbol{b}) - \operatorname{div}\boldsymbol{\sigma}\} - \{\rho(\boldsymbol{a} - \boldsymbol{b}) - \operatorname{div}\boldsymbol{\sigma}\} \otimes \boldsymbol{x} + \boldsymbol{\sigma}^{\mathrm{T}} - \boldsymbol{\sigma}]\,\mathrm{d}v = \mathbf{O}.$$

By virtue of the smoothness hypotheses which have been made concerning the motion, the density, the stress and the body force, the integrands in the transformed equations are continuous in the current configuration B_t. Furthermore, the range of integration is an arbitrary regular subregion of B_t. By appeal to Problem 1.19 (p. 42), the integrands are therefore zero and we arrive at *Cauchy's equations of motion*

$$\rho \boldsymbol{a} = \operatorname{div} \boldsymbol{\sigma} + \rho \boldsymbol{b}, \tag{23}$$

$$\boldsymbol{\sigma}^{\mathrm{T}} = \boldsymbol{\sigma}. \tag{24}$$

The partial differential equation (23) is customarily referred to as the *equation of motion*. The purely algebraic symmetry condition (24), which may be regarded as a local expression of the balance of angular momentum, implies that $\boldsymbol{\sigma}$ has three proper numbers σ_1, σ_2, σ_3; they are called the *principal stresses* and an associated set of orthonormal proper vectors defines the *principal axes of stress*.

Equation (23) is set in the spatial description. It may readily be transferred to the referential description by means of the formula $(2.19)_2$ which, in combination with equations (6) and $(20)_2$, gives

$$\operatorname{Div} \boldsymbol{s} = J \operatorname{div}(J^{-1}\boldsymbol{F}\boldsymbol{s}) = (\rho_r/\rho) \operatorname{div} \boldsymbol{\sigma}.$$

Thus, recalling also equation $(2.12)_4$, we obtain the referential form

$$\rho_r \ddot{\boldsymbol{x}} = \operatorname{Div} \boldsymbol{s} + \rho_r \boldsymbol{b}$$

of the equation of motion. From the definition $(22)_1$ it is seen that equation (24) assures the symmetry of the second Piola–Kirchhoff stress $\boldsymbol{t}$. But, as equation $(20)_1$ makes clear, the nominal stress $\boldsymbol{s}$ is not generally symmetric. Alternative referential forms of equation (24) are hence

$$\boldsymbol{s}^{\mathrm{T}}\boldsymbol{F}^{\mathrm{T}} = \boldsymbol{F}\boldsymbol{s} \quad \text{and} \quad \boldsymbol{t}^{\mathrm{T}} = \boldsymbol{t}.$$

Problem 7 An *incompressible ideal fluid* is an incompressible material of uniform density in which the stress is necessarily spherical. A material body composed of such a fluid is subjected to the motion considered in Problem 2.8 (p. 64). Assuming that no body forces act, write down the equation of motion. Deduce that the radius of the cavity at time t satisfies the differential equation

$$\rho(a\ddot{a} + \tfrac{3}{2}\dot{a}^2) = P - p_0, \tag{A}$$

where ρ is the density of the fluid, p_0 the uniform pressure in it prior to the explosion, and P the pressure exerted on the surface of the cavity by the explosion products. Given that P is proportional to the inverse square of the volume of the cavity, express $\dot{a}^2$ in terms of a and discuss the general character of the motion.

Solution. When the stress is spherical equation $(18)_2$ applies and the equation of motion (23) simplifies to

$$\rho \boldsymbol{a} = -\operatorname{grad} p + \rho \boldsymbol{b}. \tag{25}$$

For the spherically symmetric motion under consideration, $\boldsymbol{b} = \mathbf{0}$ and

$$\boldsymbol{a} = \{(a^2\ddot{a} + 2a\dot{a}^2)r^{-2} - 2a^4\dot{a}^2r^{-5}\}\hat{\mathbf{r}},$$

where $r = |\boldsymbol{x}|$ and $\hat{\boldsymbol{r}} = r^{-1}\boldsymbol{x}$, $\boldsymbol{x}$ being the position of a typical point relative to the centre o of the cavity. It follows from (25) that the pressure p is a function of r and t only and that

$$\partial p/\partial r = -\rho\{(a^2\ddot{a} + 2a\dot{a}^2)r^{-2} - 2a^4\dot{a}^2r^{-5}\}.$$

This equation can be integrated to give

$$p = p_0 + \rho\{(a^2\ddot{a} + 2a\dot{a}^2)r^{-1} - \tfrac{1}{2}a^4\dot{a}^2r^{-4}\}, \tag{B}$$

use being made of the condition $p \to p_0$ as $r \to \infty$, arising from the requirement that the pressure change due to the explosion must approach zero with increasing distance from the cavity. At the surface of the cavity, $r = a$ and $p = P$, and on making these substitutions in (B) we obtain the differential equation (A).

In order to integrate (A) we change the independent variable from t to a by setting $\ddot{a} = \frac{1}{2}\,\mathrm{d}\dot{a}^2/\mathrm{d}a$. Then

$$a\frac{\mathrm{d}\dot{a}^2}{\mathrm{d}a} + 3\dot{a}^2 = 2\rho^{-1}(P - p_0),$$

and the introduction of the integrating factor a^2 leads to

$$\frac{\mathrm{d}}{\mathrm{d}a}(a^3\dot{a}^2) = 2\rho^{-1}(P - p_0)a^2.$$

At $t = 0$ the radius of the cavity is A and since the body is initially everywhere at rest, $\dot{a} = 0$. Hence

$$a^3\dot{a}^2 = 2\rho^{-1}\int_A^a (P - p_0)a^2\,\mathrm{d}a. \tag{C}$$

The stated dependence of P upon the volume of the cavity is conveniently expressed in the form $P = P_0(a/A)^{-6}$, where P_0 is the applied pressure at $t = 0$. The integral in (C) can then be evaluated and we find that

$$\dot{a}^2 = (2P_0/3\rho)\{1 - (a/A)^{-3}\}\{(a/A)^{-3} - (p_0/P_0)\}.$$

For $\dot{a}$ to be real either $1 \leqslant a/A \leqslant (P_0/p_0)^{1/3}$ or $(P_0/p_0)^{1/3} \leqslant a/A \leqslant 1$. But the latter possibility corresponds to an implosion, not an explosion, since P_0, the maximum pressure enforced at the cavity wall, must exceed the ambient pressure p_0 if the fluid is to be driven outwards. Thus the cavity radius varies between the limits A and $(P_0/p_0)^{1/3}A$ and the body as a whole performs a steady oscillation, the particles moving in unison in the sense that the velocity is everywhere directed either towards or away from o at a given instant. We leave it to the reader to verify that the period of the motion is

$$(2\rho A^2/3P_0)^{1/2}\int_{p_0/P_0}^{1}\{(1-\xi)(\xi - p_0/P_0)\}^{-1/2}\xi^{-4/3}\,d\xi.$$

Problem 8 Investigate, in terms of the principal stresses and the principal axes of stress, the extremal properties of the normal stress $\sigma_{(\boldsymbol{n})}$ and the shear stress norm $\tau_{(\boldsymbol{n})}$ at a given point x as the unit normal $\boldsymbol{n}$ varies.

Solution. Let σ_i be the principal stresses at x and n_i the components of the unit normal $\boldsymbol{n}$ relative to an orthonormal basis $\{\boldsymbol{s}_1, \boldsymbol{s}_2, \boldsymbol{s}_3\}$ which defines the principal axes of stress at x. Then the stress tensor and the stress vector at x have the representations

$$\boldsymbol{\sigma} = \sum_{r=1}^{3} \sigma_r \boldsymbol{s}_r \otimes \boldsymbol{s}_r, \quad \boldsymbol{t}_{(\boldsymbol{n})} = \sum_{r=1}^{3} \sigma_r n_r \boldsymbol{s}_r,$$

and equations (16) and $(17)_2$ yield the following expressions for the normal stress and the shear stress norm:

$$\sigma_{(\boldsymbol{n})} = \sigma_1 n_1^2 + \sigma_2 n_2^2 + \sigma_3 n_3^2, \tag{A}$$

$$\tau_{(\boldsymbol{n})} = \{\sigma_1^2 n_1^2 + \sigma_2^2 n_2^2 + \sigma_3^2 n_3^2 - (\sigma_1 n_1^2 + \sigma_2 n_2^2 + \sigma_3 n_3^2)^2\}^{1/2} \tag{B}$$

$$= \{(\sigma_2 - \sigma_3)^2 n_2^2 n_3^2 + (\sigma_3 - \sigma_1)^2 n_3^2 n_1^2 + (\sigma_1 - \sigma_2)^2 n_1^2 n_2^2\}^{1/2}. \tag{C}$$

(*a*) Suppose first that the principal stresses are distinct. Since the values of n_i are subject to the constraint $n_1^2 + n_2^2 + n_3^2 = 1$, the extremal values of $\sigma_{(n)}$ and $\tau_{(n)}$ are found by solving the equations

$$\frac{\partial}{\partial n_i}\{\sigma_{(n)} - \mu(n_1^2 + n_2^2 + n_3^2 - 1)\} = 0$$

and

$$\frac{\partial}{\partial n_i}\{\tau_{(n)} - \nu(n_1^2 + n_2^2 + n_3^2 - 1)\} = 0$$

in which μ and ν are Lagrange multipliers. On substituting from (A) and (B) and carrying out the differentiation we obtain the relations

$$(\sigma_1 - \mu)n_1 = 0, \quad (\sigma_2 - \mu)n_2 = 0, \quad (\sigma_3 - \mu)n_3 = 0, \qquad \text{(D)}$$

and

$$\left.\begin{aligned}(\sigma_1^2 - 2\sigma_1\sigma_{(n)} - 2\nu\tau_{(n)})n_1 &= 0,\\ (\sigma_2^2 - 2\sigma_2\sigma_{(n)} - 2\nu\tau_{(n)})n_2 &= 0,\\ (\sigma_3^2 - 2\sigma_3\sigma_{(n)} - 2\nu\tau_{(n)})n_3 &= 0.\end{aligned}\right\} \qquad \text{(E)}$$

It follows from (D) that two of the n_i must be zero. Thus, from (A), the extremal values of the normal stress are the principal stresses and $\sigma_{(n)} = \sigma_i$ when $\boldsymbol{n} = \pm \boldsymbol{s}_i$, that is when $\boldsymbol{n}$ is aligned with the appropriate principal axis of stress.

From (E) we infer that at least one of the n_i must be zero, since otherwise the quadratic equation $\sigma^2 - 2\sigma\sigma_{(n)} - 2\nu\tau_{(n)} = 0$ would have three roots, namely σ_1, σ_2 and σ_3. If $n_1 = 0$, $n_2 \neq 0$, $n_3 \neq 0$, equations (C), (A) and (E) give

$$\tau_{(n)} = |(\sigma_2 - \sigma_3)n_2 n_3|, \quad \sigma_{(n)} = \sigma_2 n_2^2 + \sigma_3 n_3^2 = \tfrac{1}{2}(\sigma_2 + \sigma_3),$$

and since $n_2^2 + n_3^2 = 1$ it follows that

$$n_1 = 0, \quad n_2 = \pm 1/\sqrt{2}, \quad n_3 = \pm 1/\sqrt{2}, \quad \tau_{(n)} = \tfrac{1}{2}|\sigma_2 - \sigma_3|.$$

The corresponding solutions of equations (E) in which n_2 and n_3 are in turn zero are

$$\begin{aligned} &n_1 = \pm 1/\sqrt{2}, \quad n_2 = 0, && n_3 = \pm 1/\sqrt{2}, \quad \tau_{(n)} = \tfrac{1}{2}|\sigma_3 - \sigma_1|,\\ &n_1 = \pm 1/\sqrt{2}, \quad n_2 = \pm 1/\sqrt{2}, && n_3 = 0, \quad \tau_{(n)} = \tfrac{1}{2}|\sigma_1 - \sigma_2|.\end{aligned}$$

There are further solutions of (E) in which two of the n_i are zero, the corresponding values of $\tau_{(n)}$, from (C), being zero. The extremal values of the shear stress norm are therefore half the magnitudes of the principal stress differences and zero.

(*b*) Considering next the case in which two of the principal stresses are equal, suppose that $\sigma_1 = \sigma_2 \neq \sigma_3$. Equations (A) and (C) then take on the simplified forms

$$\sigma_{(n)} = \tfrac{1}{2}(\sigma_1 + \sigma_3) - \tfrac{1}{2}(\sigma_1 - \sigma_3)\cos 2\theta, \quad \tau_{(n)} = \tfrac{1}{2}|(\sigma_3 - \sigma_1)\sin 2\theta|,$$

where $\theta = \cos^{-1} n_3$ $(0 \leqslant \theta \leqslant \pi)$ is the angle between $\boldsymbol{n}$ and the principal axis of stress associated with σ_3. The extremal values of $\sigma_{(n)}$ and $\tau_{(n)}$ are σ_1, σ_3 and $\tfrac{1}{2}|\sigma_1 - \sigma_3|$, 0 respectively, in conformity with the results obtained in (*a*).

(*c*) Finally, if $\sigma_1 = \sigma_2 = \sigma_3$, the stress at x is spherical and $\sigma_{(n)} = \sigma_1$, $\tau_{(n)} = 0$ for all orientations of $\boldsymbol{n}$.

We conclude from the foregoing discussion that the maximum shear stress norm at x is half the difference between the greatest and least principal stresses and that the corresponding normal directions are orthogonal to the principal axis associated with the intermediate principal stress and inclined at either $\tfrac{1}{4}\pi$ or $\tfrac{3}{4}\pi$ to the other principal axes.

Problem 9 Obtain the formula

$$\bar{\boldsymbol{\sigma}} = \tfrac{1}{2}V^{-1}\int_B \rho(\boldsymbol{x} \otimes \boldsymbol{b} + \boldsymbol{b} \otimes \boldsymbol{x})\, dv + \tfrac{1}{2}V^{-1}\int_{\partial B} (\boldsymbol{x} \otimes \boldsymbol{t}_{(n)} + \boldsymbol{t}_{(n)} \otimes \boldsymbol{x})\, da \qquad \text{(A)}$$

for the mean stress in a material body which is permanently at rest in the regular configuration B, V being the volume of B.

If the boundary ∂B consists of two closed surfaces ∂B_1 and ∂B_2, the former enclosing the latter, and uniform pressures p_1 and p_2 act on ∂B_1 and ∂B_2 respectively, show that, in the absence of body forces,

$$\bar{\boldsymbol{\sigma}} = -(V_1 - V_2)^{-1}(p_1 V_1 - p_2 V_2)\boldsymbol{I},$$

where V_1 and V_2 are the volumes enclosed by ∂B_1 and ∂B_2 respectively.

Solution. For a material body which is permanently at rest the acceleration is zero and the equation of motion (23) reduces to the *equilibrium equation*

$$\operatorname{div} \boldsymbol{\sigma} + \rho \boldsymbol{b} = \mathbf{0}. \tag{26}$$

Thus, for the given body,

$$\int_B \{\boldsymbol{x} \otimes (\operatorname{div} \boldsymbol{\sigma} + \rho \boldsymbol{b}) + (\operatorname{div} \boldsymbol{\sigma} + \rho \boldsymbol{b}) \otimes \boldsymbol{x}\} \, dv = \mathbf{O}. \tag{B}$$

With the aid of the divergence theorem, in the form (1.95), (B) can be transformed into

$$\int_B \{\rho(\boldsymbol{x} \otimes \boldsymbol{b} + \boldsymbol{b} \otimes \boldsymbol{x}) - \boldsymbol{\sigma} - \boldsymbol{\sigma}^{\mathrm{T}}\} \, dv + \int_{\partial B} \{\boldsymbol{x} \otimes (\boldsymbol{\sigma}^{\mathrm{T}} \boldsymbol{n}) + (\boldsymbol{\sigma}^{\mathrm{T}} \boldsymbol{n}) \otimes \boldsymbol{x}\} \, da = \mathbf{O}.$$

The required formula (A) is obtained on using equations (15) and (24) and noting that the mean stress in B is defined by

$$\bar{\boldsymbol{\sigma}} = V^{-1} \int_B \boldsymbol{\sigma} \, dv. \tag{C}$$

[Equation (A) shows that the mean stress in a material body in equilibrium depends only upon the body forces and the boundary tractions; no information about the internal state of stress is needed. The reader may have noticed that the omission of the second tensor product from (B) leads to the seemingly simpler result

$$\bar{\boldsymbol{\sigma}} = V^{-1} \int_B \rho \boldsymbol{x} \otimes \boldsymbol{b} \, dv + V^{-1} \int_{\partial B} \boldsymbol{x} \otimes \boldsymbol{t}_{(\boldsymbol{n})} \, da. \tag{D}$$

But, in contrast to (A) and (C), this expression for $\bar{\boldsymbol{\sigma}}$ is not necessarily symmetric. What is the reason for the apparent discrepancy?]

In the special case illustrated in Figure 4, $\boldsymbol{b} = \mathbf{0}$ and the boundary tractions are given by

$$\boldsymbol{t}_{(\boldsymbol{n})} = -p_1 \boldsymbol{n}_1 \quad \text{on} \quad \partial B_1, \quad \boldsymbol{t}_{(\boldsymbol{n})} = -p_2 \boldsymbol{n}_2 \quad \text{on} \quad \partial B_2,$$

the directions of the unit normals $\boldsymbol{n}_1$ and $\boldsymbol{n}_2$ being as shown. Hence, from equation (A),

$$\bar{\boldsymbol{\sigma}} = -\tfrac{1}{2}(V_1 - V_2)^{-1} \left(p_1 \int_{\partial B_1} (\boldsymbol{x} \otimes \boldsymbol{n}_1 + \boldsymbol{n}_1 \otimes \boldsymbol{x}) \, da + p_2 \int_{\partial B_2} (\boldsymbol{x} \otimes \boldsymbol{n}_2 + \boldsymbol{n}_2 \otimes \boldsymbol{x}) \, da \right),$$

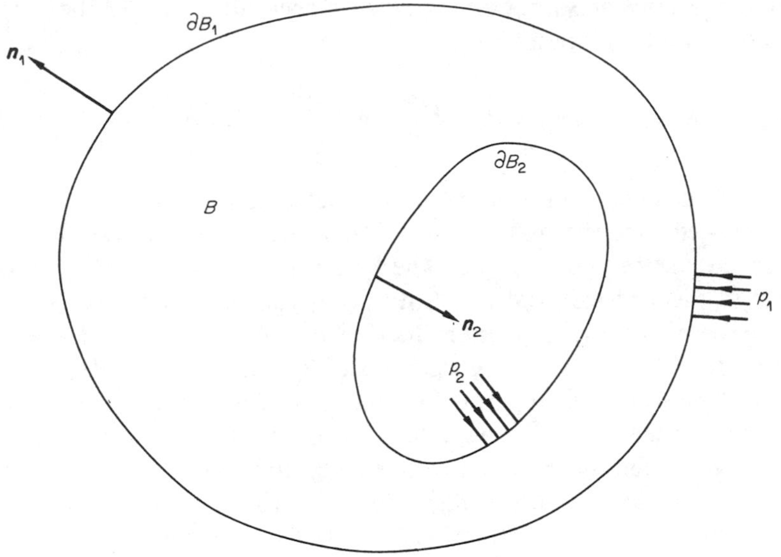

FIGURE 4 Surface tractions and outward normals (Problem 9).

the volume of B being $V_1 - V_2$. The divergence theorem (1.97) provides the identity

$$\int_{\partial B_1} (\boldsymbol{x} \otimes \boldsymbol{n}_1 + \boldsymbol{n}_1 \otimes \boldsymbol{x})\,\mathrm{d}a = \int_{B_1} \{\operatorname{grad} \boldsymbol{x} + (\operatorname{grad} \boldsymbol{x})^{\mathrm{T}}\}\,\mathrm{d}v$$
$$= 2\boldsymbol{I} \int_{B_1} \mathrm{d}v = 2V_1\boldsymbol{I},$$

and similarly,

$$\int_{\partial B_2} (\boldsymbol{x} \otimes \boldsymbol{n}_2 + \boldsymbol{n}_2 \otimes \boldsymbol{x})\,\mathrm{d}a = -2V_2\boldsymbol{I},$$

the sign being reversed here because $\boldsymbol{n}_2$ is directed *inwards* to the region bounded by ∂B_2. The stated result now follows.

5 ENERGY

The *kinetic energy*, $K(R_t)$, of the material occupying R_t is defined by

$$K(R_t) = \int_{R_t} \tfrac{1}{2}\rho \boldsymbol{v} \cdot \boldsymbol{v}\,\mathrm{d}v, \tag{27}$$

and the *mechanical power* (or *rate-of-working*), $P(R_t)$, of the forces acting on this material by

$$P(R_t) = \int_{R_t} \rho \boldsymbol{b} \cdot \boldsymbol{v} \, \mathrm{d}v + \int_{\partial R_t} \boldsymbol{t}_{(\boldsymbol{n})} \cdot \boldsymbol{v} \, \mathrm{d}a. \tag{28}$$

The same notation is used here as in Sections 1 and 2 and it is again supposed that the material under consideration is non-polar and not subject to body torques. The balance of energy in the representative piece of material currently occupying R_t equates the rate of change of the energy content to the sum of the integrated rates of supply of energy to volume elements of R_t and on surface elements of ∂R_t. The rate of change of the kinetic energy and the mechanical power appear on opposite sides of this equation, but they do not account in full for the balance of energy, first because a deformable body does not possess energy solely by virtue of its motion, and secondly because the interconvertibility of mechanical work and heat which is a fundamental physical principle of energy transfer would be excluded. It is therefore postulated that in each of the configurations $\{B_t\}$ of a moving material body there are associated with an arbitrary material region R_t scalars $U(R_t)$ and $H(R_t)$, referred to respectively as the *internal energy content* and the *heating* of the material occupying R_t, such that

$$\frac{\mathrm{d}}{\mathrm{d}t}\{K(R_t) + U(R_t)\} = P(R_t) + H(R_t). \tag{29}$$

This is the *equation of energy balance* and the set functions U and H represent in turn the energy which the material possesses in addition to its kinetic energy and the rate at which energy is supplied to the material as a result of non-mechanical interaction with its surroundings. The internal energy U is assumed to have the properties (i) and (ii) ascribed to the mass function in Section 1. These conditions ensure the existence of a scalar field ε such that

$$U(R_t) = \int_{R_t} \rho \varepsilon \, \mathrm{d}v: \tag{30}$$

ε is referred to as the *internal energy* (per unit mass) of the material. Further, by analogy with equations (10), the heating is taken to be expressible in the form

$$H(R_t) = \int_{R_t} \rho r \, \mathrm{d}v + \int_{\partial R_t} h_{(\boldsymbol{n})} \, \mathrm{d}a. \tag{31}$$

The scalar field r specifies the rate per unit mass at which heat is communicated to the material and is known as the *heat supply*. An example is the heating caused by a distribution of radioactive sources. The *heat flux* $h_{(\boldsymbol{n})}$ measures the rate per unit area at which heat is imparted to the material occupying R_t across the boundary ∂R_t. The scalars ε, r and $h_{(\boldsymbol{n})}$ are defined on the same domains as ρ, $\boldsymbol{b}$ and $\boldsymbol{t}_{(\boldsymbol{n})}$ respectively *and they are assumed to share with their counterparts the smoothness properties stated previously.*[4]

By entering the expressions (27), (28), (30) and (31) into (29) we can now present the equation of energy balance in integral form as

$$\frac{\mathrm{d}}{\mathrm{d}t}\int_{R_t} \rho(\tfrac{1}{2}\boldsymbol{v}\cdot\boldsymbol{v} + \varepsilon)\,\mathrm{d}v = \int_{R_t} \rho(\boldsymbol{b}\cdot\boldsymbol{v} + r)\,\mathrm{d}v + \int_{\partial R_t} (\boldsymbol{t}_{(\boldsymbol{n})}\cdot\boldsymbol{v} + h_{(\boldsymbol{n})})\,\mathrm{d}a. \tag{32}$$

Problem 10 Obtain the *equations of mechanical* and *thermal energy balance,*

$$\frac{\mathrm{d}}{\mathrm{d}t} K(R_t) + S(R_t) = P(R_t) \quad \text{and} \quad \frac{\mathrm{d}}{\mathrm{d}t} U(R_t) - S(R_t) = H(R_t), \tag{33}$$

where

$$S(R_t) = \int_{R_t} \operatorname{tr}(\boldsymbol{\sigma}\boldsymbol{D})\,\mathrm{d}v. \tag{34}$$

Solution. Either of equations (33) implies the other through the overall energy balance (29). It therefore suffices to derive $(33)_1$ and this is done by first forming the scalar product of each term of the equation of motion (23) with the velocity, then integrating the resulting equation term-by-term over the arbitrary regular material region R_t. The first step yields

$$\rho\boldsymbol{a}\cdot\boldsymbol{v} = (\operatorname{div}\boldsymbol{\sigma})\cdot\boldsymbol{v} + \rho\boldsymbol{b}\cdot\boldsymbol{v}$$

which, with the use of equation $(2.12)_2$ and Exercise 1.14 (ii) (p. 49), can

[4] See the italicized statements on pp. 88, 92 and 93.

be rewritten as

$$\rho(\tfrac{1}{2}\boldsymbol{v}\cdot\boldsymbol{v})^{\boldsymbol{\cdot}} + \operatorname{tr}(\boldsymbol{\sigma L}) = \operatorname{div}(\boldsymbol{\sigma v}) + \rho\boldsymbol{b}\cdot\boldsymbol{v}.$$

Also, recalling the definition (2.43) and Exercise 1.8 (ii) (p. 48), we have

$$\operatorname{tr}(\boldsymbol{\sigma L}) = \operatorname{tr}(\boldsymbol{\sigma D}) + \operatorname{tr}(\boldsymbol{\sigma W}) = \operatorname{tr}(\boldsymbol{\sigma D}).$$

Hence, proceeding to the integration,

$$\int_{R_t} \rho(\tfrac{1}{2}\boldsymbol{v}\cdot\boldsymbol{v})^{\boldsymbol{\cdot}}\,dv + \int_{R_t} \operatorname{tr}(\boldsymbol{\sigma D})\,dv = \int_{R_t} \rho\boldsymbol{b}\cdot\boldsymbol{v}\,dv + \int_{R_t} \operatorname{div}(\boldsymbol{\sigma v})\,dv. \quad \text{(A)}$$

Equation $(33)_1$ is now reached on applying the scalar formula (4) to the first term on the left of (A) and the divergence theorem, in the form (1.98), to the second term on the right. The details of the latter step are

$$\int_{R_t} \operatorname{div}(\boldsymbol{\sigma v})\,dv = \int_{\partial R_t} \boldsymbol{n}\cdot(\boldsymbol{\sigma v})\,da = \int_{\partial R_t} \boldsymbol{v}\cdot(\boldsymbol{\sigma}^{\mathrm{T}}\boldsymbol{n})\,da = \int_{\partial R_t} \boldsymbol{t}_{(\boldsymbol{n})}\cdot\boldsymbol{v}\,da.$$

The scalar field $\operatorname{tr}(\boldsymbol{\sigma D})$ is called the *stress power* per unit volume in B_t. On referring the tensors $\boldsymbol{\sigma}$ and $\boldsymbol{D}$ to the orthonormal basis $\{\boldsymbol{r}_1, \boldsymbol{r}_2, \boldsymbol{r}_3\}$ composed of vectors defining the principal axes of stretching we obtain the representation

$$\operatorname{tr}(\boldsymbol{\sigma D}) = \sum_{p=1}^{3} \{\boldsymbol{r}_p\cdot(\boldsymbol{\sigma r}_p)\}\nu_p = \sigma_{(\boldsymbol{r}_p)}\nu_p$$

from which the stress power at a point x is seen to be the sum of the products of the principal stretchings ν_i at x with the normal stresses $\sigma_{(\boldsymbol{r}_i)}$ on surface elements at x orthogonal to the principal axes of stretching. This intepretation makes explicit the sense in which the stress power may be viewed as the rate of working of the stress occasioned by the stretching of volume elements. Equations (33) show that the integrated stress power S effectively couples together the mechanical and thermal energy balances. The rate of change of the internal energy content of the material occupying R_t may be regarded as having a mechanical component $S(R_t)$ and a thermal component $(d/dt)U(R_t) - S(R_t)$. In general, however, $S(R_t)$ cannot be expressed as the rate of change of the integral over R_t of a scalar field.

We observe in passing that when the stress is spherical thc stress power is given by $-p \operatorname{tr} \boldsymbol{D} = -p \operatorname{div} \boldsymbol{v}$. In an incompressible ideal fluid the stress power therefore vanishes identically and the mechanical and thermal energy balances are uncoupled. This means that the motion and the thermal state of the fluid can have no influence on one another.

Problem 11 Obtain expressions for the stress power in terms of the nominal stress $\boldsymbol{s}$ and the second Piola–Kirchhoff stress $\boldsymbol{t}$.

Solution. The reader is first reminded of the general properties (1.31) and (1.53) of the trace operator.

(*a*)

$$\begin{aligned} \operatorname{tr}(\boldsymbol{\sigma D}) &= \operatorname{tr}(\boldsymbol{\sigma L}) = \operatorname{tr}(\boldsymbol{L\sigma}) = J^{-1}\operatorname{tr}(\boldsymbol{LFs}) = J^{-1}\operatorname{tr}(\dot{\boldsymbol{F}}\boldsymbol{s}) \\ &= (\rho/\rho_r)\operatorname{tr}(\boldsymbol{s}\dot{\boldsymbol{F}}). \end{aligned} \tag{35}$$

Here successive use has been made of equations $(20)_2$, $(2.30)_1$ and (6).

(*b*)

$$\begin{aligned} \operatorname{tr}(\boldsymbol{\sigma D}) &= \operatorname{tr}(\boldsymbol{\sigma L}) = \tfrac{1}{2}\operatorname{tr}(\boldsymbol{L\sigma} + \boldsymbol{\sigma L}^{\mathrm{T}}) = \tfrac{1}{2}J^{-1}\operatorname{tr}(\boldsymbol{LFtF}^{\mathrm{T}} + \boldsymbol{FtF}^{\mathrm{T}}\boldsymbol{L}^{\mathrm{T}}) \\ &= \tfrac{1}{2}J^{-1}\operatorname{tr}(\dot{\boldsymbol{F}}\boldsymbol{tF}^{\mathrm{T}} + \boldsymbol{Ft}\dot{\boldsymbol{F}}^{\mathrm{T}}) = \tfrac{1}{2}J^{-1}\operatorname{tr}\{(\boldsymbol{F}^{\mathrm{T}}\dot{\boldsymbol{F}} + \dot{\boldsymbol{F}}^{\mathrm{T}}\boldsymbol{F})\boldsymbol{t}\} \\ &= \tfrac{1}{2}J^{-1}\operatorname{tr}(\dot{\boldsymbol{C}}\boldsymbol{t}) = \tfrac{1}{2}(\rho/\rho_r)\operatorname{tr}(\boldsymbol{t}\dot{\boldsymbol{C}}), \end{aligned}$$

the steps being justified by equations $(22)_2$, $(2.30)_1$, $(2.24)_1$ and (6).

The application of the equation of thermal energy balance $(33)_2$ to a tetrahedral subregion of B_t establishes the existence of a *heat flux vector* $\boldsymbol{q}$ through which the dependence of the heat flux $h_{(\boldsymbol{n})}$ upon the unit normal $\boldsymbol{n}$ is given by

$$h_{(\boldsymbol{n})} = \boldsymbol{q}\cdot\boldsymbol{n}. \tag{36}$$

This result occupies a position in the analysis of heat flow analogous to that of equation (15) in the theory of stress. Like the stress tensor, $\boldsymbol{q}$ is defined on the configurations $\{B_t\}$ *and it is assumed to have the degree of smoothness attributed to* $\boldsymbol{\sigma}$ *in Section 4.*[5] It follows from (36) that, with reference to a rectangular Cartesian system (o, e) of spatial coordinates and a representative point x in B_t, $q_i = \boldsymbol{q}\cdot\boldsymbol{e}_i$

[5] See the italicized statement on p. 100.

is the *influx* of heat per unit area in B_t on a surface segment whose outward normal at x is in the i-direction.

The *referential heat flux vector*, $\boldsymbol{Q}$, is defined by

$$\boldsymbol{Q} = J\boldsymbol{F}^{-1}\boldsymbol{q}, \quad \boldsymbol{q} = J^{-1}\boldsymbol{F}\boldsymbol{Q}, \tag{37}$$

and the formula

$$\boldsymbol{q} \cdot \boldsymbol{n}\, \mathrm{d}a = \boldsymbol{Q} \cdot \boldsymbol{N}\, \mathrm{d}A,$$

corresponding to (21), is obtained with the aid of (2.21). In relation to a rectangular Cartesian system $(O, \boldsymbol{E})$ of referential coordinates, $Q_\alpha = \boldsymbol{Q} \cdot \boldsymbol{E}_\alpha$ is therefore the influx of heat per unit area in the reference configuration B_r on a material surface segment through x whose configuration in B_r has its outward normal in the α-direction.

On substituting from equations (30), (31) and (34) into $(33)_2$ we can write the equation of thermal energy balance in the integral form

$$\frac{\mathrm{d}}{\mathrm{d}t} \int_{R_t} \rho \varepsilon \, \mathrm{d}v = \int_{R_t} \{\rho r + \mathrm{tr}\,(\boldsymbol{\sigma}\boldsymbol{D})\}\, \mathrm{d}v + \int_{\partial R_t} \boldsymbol{q} \cdot \boldsymbol{n}\, \mathrm{d}a, \tag{38}$$

use being made also of (36). When the first and last terms in (38) are transformed by means of equation (4) and the divergence theorem (1.98) we obtain

$$\int_{R_t} \{\rho \dot{\varepsilon} - \mathrm{tr}\,(\boldsymbol{\sigma}\boldsymbol{D}) - \mathrm{div}\, \boldsymbol{q} - \rho r\}\, \mathrm{d}v = 0,$$

and in view of the continuity in B_t of the integrand and the arbitrariness of the regular material region R_t there follows the *energy equation*

$$\rho \dot{\varepsilon} = \mathrm{tr}\,(\boldsymbol{\sigma}\boldsymbol{D}) + \mathrm{div}\, \boldsymbol{q} + \rho r. \tag{39}$$

This is the field equation associated with the balance law (32). It can be transferred from the spatial description in which it has been derived to the referential description with the aid of equations $(2.19)_1$ and (35). The first of these identities, in conjunction with $(37)_2$ and (6), gives

$$\mathrm{Div}\, \boldsymbol{Q} = J\, \mathrm{div}\, (J^{-1} \boldsymbol{F}\boldsymbol{Q}) = (\rho_r/\rho)\, \mathrm{div}\, \boldsymbol{q},$$

and the referential counterpart of equation (39) therefore emerges as

$$\rho_r \dot{\varepsilon} = \mathrm{tr}\,(\boldsymbol{s}\dot{\boldsymbol{F}}) + \mathrm{Div}\, \boldsymbol{Q} + \rho_r r.$$

Problem 12 Show that, for a material body not subject to body forces or heat supply, the rate of change of the total energy (i.e. kinetic energy plus internal energy content) of the material occupying an arbitrary regular region R_t is equal to the total efflux across the boundary ∂R_t of the vector $\boldsymbol{j} = -\boldsymbol{\sigma}\boldsymbol{v} - \boldsymbol{q}$.

Solution. Under the given conditions the volume integral on the right-hand side of the equation of energy balance (32) vanishes. And with the aid of equations (15) and (36) the integrand of the surface integral can be put into the form

$$\boldsymbol{v}\,.\,(\boldsymbol{\sigma}^{\mathrm{T}}\boldsymbol{n}) + \boldsymbol{q}\,.\,\boldsymbol{n} = \boldsymbol{n}\,.\,(\boldsymbol{\sigma}\boldsymbol{v} + \boldsymbol{q}) = -\,\boldsymbol{j}\,.\,\boldsymbol{n}.$$

Thus (32) reduces to

$$\frac{\mathrm{d}}{\mathrm{d}t}\int_{R_t} \rho(\tfrac{1}{2}\boldsymbol{v}\,.\,\boldsymbol{v} + \varepsilon)\,\mathrm{d}v = -\int_{\partial R_t} \boldsymbol{j}\,.\,\boldsymbol{n}\,\mathrm{d}a$$

from which it is clear that $\boldsymbol{j}$ has the stated property.

[In view of this result $\boldsymbol{j}$ is known as the *energy flux vector.*]

The properties which characterize an incompressible ideal fluid have been stated in Problem 7 (p. 101). A *compressible ideal fluid* is a material in which the stress is necessarily spherical and the heat flux vector identically zero. A motion of such a fluid in which the pressure and the internal energy are entirely determined by the density is said to be *barotropic*.

Problem 13 A *perfect gas* is a compressible ideal fluid in which, in the absence of heat supply, the motion is necessarily barotropic, the pressure being proportional to ρ^{γ} where $\gamma(>1)$ is a constant. Show that when $r = 0$ the internal energy of a perfect gas is given by

$$\varepsilon = (\gamma - 1)^{-1}(p/\rho) + \text{constant}. \tag{40}$$

Solution. When the stress is spherical and the heat flux vector and heat supply are zero the energy equation (39) reduces to

$$\rho\dot{\varepsilon} + p\,\mathrm{div}\,\boldsymbol{v} = 0.$$

In a barotropic motion,

$$\dot{\varepsilon} = (\mathrm{d}\varepsilon/\mathrm{d}\rho)\dot{\rho} = -\rho(\mathrm{d}\varepsilon/\mathrm{d}\rho)\,\mathrm{div}\,\boldsymbol{v},$$

use being made of the spatial continuity equation (3). Hence, for a perfect gas not subject to heat supply,

$$\{\rho^2(\mathrm{d}\varepsilon/\mathrm{d}\rho) - p\} \operatorname{div} \boldsymbol{v} = 0. \tag{A}$$

In this equation the contents of the curly brackets depend only upon ρ and since the result holds for all motions of the gas it must follow[6] that $\rho^2(\mathrm{d}\varepsilon/\mathrm{d}\rho) = p$. On setting $p = k\rho^{\gamma}$ in this differential equation (k being a constant) and integrating, we obtain the required expression for ε.

6 JUMP CONDITIONS

In the derivation of the field equations (3), (23), (24) and (39) the motion of the material body $\mathscr{B}$ and the fields $\rho, \boldsymbol{\sigma}, \varepsilon$ and $\boldsymbol{q}$, representing properties of the material composing $\mathscr{B}$, are assumed to satisfy certain smoothness requirements.[7] Problems of considerable physical interest arise in which these conditions are not met and particular importance attaches to situations in which the failure of the fields to attain the requisite degree of smoothness is confined to one or more surfaces. We now consider the forms taken by the balance laws (2), (12) and (32) on such *singular surfaces*.

In a motion of $\mathscr{B}$ represented by equations (2.3) and (2.4) a singular surface Σ *of order* n ($\geqslant 1$) is an orientable smooth surface on which the functions $\boldsymbol{\psi}$ and $\boldsymbol{\Psi}$ together with their derivatives of orders $1, 2, \ldots, n-1$ with respect to position and time are continuous and the derivatives of order n almost everywhere discontinuous. If $\boldsymbol{n}$ is a unit normal vector field on Σ the difference between the values taken by a field ϕ on the sides of Σ towards which and away from which $\boldsymbol{n}$ is directed is called the *jump* of ϕ on Σ and is denoted by $[\phi]$. A singular surface which propagates relative to the material is sometimes referred to as a *wave* and the term *contact discontinuity* is applied to a material singular surface. When Σ is a wave the normal

[6] A strict proof of this assertion proceeds by contradiction. Suppose that there is a value of ρ, $\bar{\rho}$ say, for which $\rho^2(\mathrm{d}\varepsilon/\mathrm{d}\rho) - p \neq 0$, and consider the motion given by

$$\boldsymbol{x} = (\rho_r/\bar{\rho})^{\frac{1}{3}}\{1 + \tfrac{1}{3}\nu(t - \bar{t})\}\boldsymbol{X},$$

where ν and $\bar{t}$ are positive constants. A straightforward calculation, which is left to the reader, shows that at $t = \bar{t}$, ρ and $\operatorname{div} \boldsymbol{v}$ have the uniform values $\bar{\rho}$ and ν respectively which means that (A) is violated. No value of ρ with the stated property can therefore exist.

[7] See the italicized statements on pp. 51, 88, 100, 109 and 111.

vector $\boldsymbol{n}$ is taken to be in the direction of propagation.

In order to derive the discontinuity relations which hold on a singular surface on account of the balance laws we must first extend the transport formulae (2.37) and (2.40) to the case in which the region R_t is intersected by a singular surface on which the fields ϕ and $\boldsymbol{u}$ are discontinuous. The alternative forms of these results stated in Exercise 2.6 (p. 85) form a convenient starting point and we notice that the velocity field $\boldsymbol{v}$ enters the formulae only through its normal component on the boundary ∂R_t.

At time t let R_t^+, ∂R_t^+ and R_t^-, ∂R_t^- denote the parts of R_t and ∂R_t situated ahead of and to the rear of the singular surface Σ as shown in Figure 5, and let $\Lambda = \Sigma \cap R_t$. In general the regions $N^\pm = R_t^\pm \cup \Lambda$

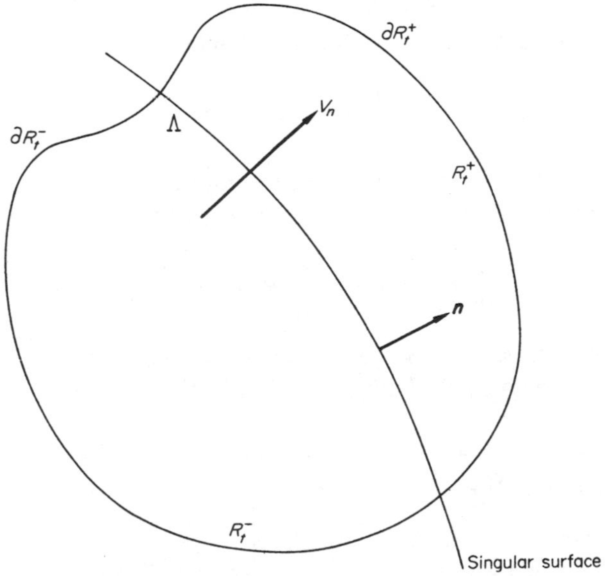

FIGURE 5 Material region intersected by a singular surface.

are not material with respect to the motion of $\mathscr{B}$, but they may be regarded as material regions in a fictitious motion in which the velocity is equal to $\boldsymbol{v}$ on the portions $\partial R_t^\pm$ of the boundaries of $N^\pm$ and equal to $V_n\boldsymbol{n}$ on the common part Λ, V_n being the speed of propagation of Σ. The application of the scalar transport formula over

N^+ and N^- individually then gives

$$\frac{\mathrm{d}}{\mathrm{d}t}\int_{N^+} \phi \,\mathrm{d}v = \int_{N^+} \frac{\partial \phi}{\partial t}\,\mathrm{d}v + \int_{\partial R_t^+} \phi \boldsymbol{v}\,.\,\boldsymbol{n}\,\mathrm{d}a - \int_A \phi^+ V_n\,\mathrm{d}a,$$

$$\frac{\mathrm{d}}{\mathrm{d}t}\int_{N^-} \phi \,\mathrm{d}v = \int_{N^-} \frac{\partial \phi}{\partial t}\,\mathrm{d}v + \int_{\partial R_t^-} \phi \boldsymbol{v}\,.\,\boldsymbol{n}\,\mathrm{d}a + \int_A \phi^- V_n\,\mathrm{d}a,$$

where the superscripts + and − in the final terms indicate evaluation on the forward and rearward sides of Σ respectively, and it is understood that the rates of change on the left-hand sides are in relation to the fictitious motion. When these equations are added the required generalization of equation (2.37) is obtained in the form[8]

$$\frac{\mathrm{d}}{\mathrm{d}t}\int_{R_t} \phi \,\mathrm{d}v = \int_{R_t} \frac{\partial \phi}{\partial t}\,\mathrm{d}v + \int_{\partial R_t} \phi \boldsymbol{v}\,.\,\boldsymbol{n}\,\mathrm{d}a - \int_A [\phi]\, V_n\,\mathrm{d}a, \qquad (41)$$

and the corresponding modification of (2.40) is

$$\frac{\mathrm{d}}{\mathrm{d}t}\int_{R_t} \boldsymbol{u} \,\mathrm{d}v = \int_{R_t} \frac{\partial \boldsymbol{u}}{\partial t}\,\mathrm{d}v + \int_{\partial R_t} \boldsymbol{u}(\boldsymbol{v}\,.\,\boldsymbol{n})\,\mathrm{d}a - \int_A [\boldsymbol{u}] V_n\,\mathrm{d}a. \qquad (42)$$

The results which have just been obtained are now applied to the balance laws (2), (12) and (32) which we write in the generic form

$$\frac{\mathrm{d}}{\mathrm{d}t}\int_{R_t} \rho\pi \,\mathrm{d}v = \int_{R_t} \rho s\,\mathrm{d}v + \int_{\partial R_t} f_{(\boldsymbol{n})}\,\mathrm{d}a, \qquad (43)$$

the quantities π, s (the supply of π per unit mass) and $f_{(\boldsymbol{n})}$ (the influx of π per unit area in B_t) being specified in the accompanying table. In

Balance law	π	s	$f_{(\boldsymbol{n})}$
Mass	1	0	0
Linear momentum	$\boldsymbol{v}$	$\boldsymbol{b}$	$\boldsymbol{t}_{(\boldsymbol{n})}$
Angular momentum	$\boldsymbol{x}\wedge\boldsymbol{v}$	$\boldsymbol{x}\wedge\boldsymbol{b}$	$\boldsymbol{x}\wedge\boldsymbol{t}_{(\boldsymbol{n})}$
Energy	$\varepsilon + \frac{1}{2}\boldsymbol{v}\,.\,\boldsymbol{v}$	$\boldsymbol{b}\,.\,\boldsymbol{v} + r$	$\boldsymbol{t}_{(\boldsymbol{n})}\,.\,\boldsymbol{v} + h_{(\boldsymbol{n})}$

[8] In connection with the addition of the left-hand sides, note that the restrictions to the boundary of $R_t (= N^+ \cup N^-)$ of the fictitious and actual motions are identical.

combination with (41) or (42), equation (43) yields

$$\int_{R_t}\left(\frac{\partial}{\partial t}(\rho\pi) - \rho s\right)\mathrm{d}v + \int_{\partial R_t}(\rho\pi\boldsymbol{v}\,.\,\boldsymbol{n} - f_{(\boldsymbol{n})})\,\mathrm{d}a - \int_{\Lambda}[\rho\pi]V_n\,\mathrm{d}a = 0,$$

whereupon the volume of R_t is made to approach zero in such a way that, in the limit, ∂R_t collapses onto the two sides of Λ as illustrated in Figure 6. When ρ, π and s are assumed to retain away from Σ

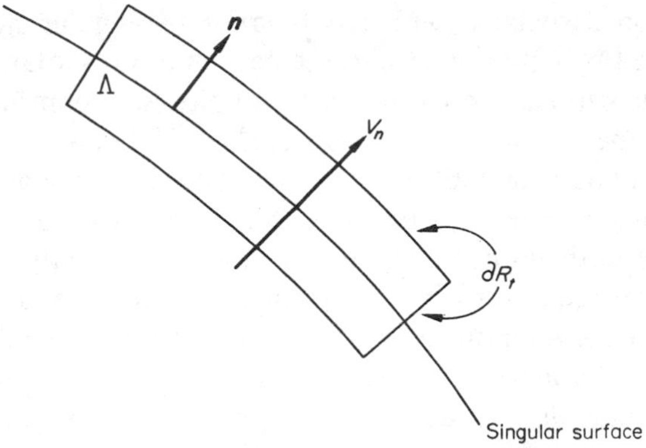

FIGURE 6 Collapse of the surface of a material region onto the two sides of a singular surface.

the smoothness properties attributed to them previously[9] the result of this procedure is[10]

$$\int_{\Lambda}([\rho\pi\boldsymbol{v}\,.\,\boldsymbol{n} - f_{(\boldsymbol{n})}] - [\rho\pi]V_n)\,\mathrm{d}a = 0. \tag{44}$$

Equation (44) holds for all segments of Σ formed by intersection with regular regions of B_t. Thus, provided that the integrand is continuous on Σ, there follows[11] the general jump condition

$$[\rho V\pi + f_{(\boldsymbol{n})}] = 0, \tag{45}$$

where

$$V = V_n - \boldsymbol{v}\,.\,\boldsymbol{n}, \tag{46}$$

[9] See the italicized statements on pp. 51, 88, 92 and 109.
[10] Note that, in view of equations (15) and (36), $f_{(-\boldsymbol{n})} = -f_{(\boldsymbol{n})}$.
[11] Once again we refer to the remarks in parentheses at the end of Problem 1.19 (p. 42).

known as the *local speed of propagation* of Σ, is the speed of travel of the singular surface *relative to the material.* Substitution into (45) of the entries in the second and fourth columns of the table leads at once to the *basic jump conditions*

$$[\rho V] = 0, \tag{47}$$

$$[\rho V \boldsymbol{v} + \boldsymbol{t}_{(\boldsymbol{n})}] = \boldsymbol{0}, \tag{48}$$

$$[\rho V(\varepsilon + \tfrac{1}{2}\boldsymbol{v} \cdot \boldsymbol{v}) + \boldsymbol{t}_{(\boldsymbol{n})} \cdot \boldsymbol{v} + h_{(\boldsymbol{n})}] = 0, \tag{49}$$

the condition associated with the balance of angular momentum being implied by equation (48) in the case of a non-polar material.

A singular surface Σ on which the velocity is discontinuous is of order 1 in the classification given earlier. If $[\boldsymbol{v} \cdot \boldsymbol{n}] \neq 0$ on Σ it follows from equation (46) that the local speed of propagation V is discontinuous and hence, from (47), that Σ is in motion relative to the material both immediately ahead of it and directly to its rear. A first-order singular surface is accordingly referred to as a *shock wave* when it carries a jump in the normal component of velocity. When the velocity jump on Σ is entirely tangential (i.e. $[\boldsymbol{v} \cdot \boldsymbol{n}] = 0, [\boldsymbol{v} \wedge \boldsymbol{n}] \neq \boldsymbol{0}$), the term *vortex sheet* is used. On a second-order singular surface the velocity and the deformation gradient are continuous. In consequence of (46) and the material continuity equation (6) the jump condition (47) is then satisfied identically and equations (48) and (49) take on the simplified forms

$$[\boldsymbol{t}_{(\boldsymbol{n})}] = \boldsymbol{0}, \quad \rho V[\varepsilon] + [h_{(\boldsymbol{n})}] = 0.$$

A propagating second-order singular surface is known as an *acceleration wave.*

Problem 14 Show that in a material in which the stress is spherical a vortex sheet is necessarily a contact discontinuity. If the material is a compressible ideal fluid show that the basic jump conditions on a propagating singular surface can be expressed as

$$[\rho V] = 0, \ [p + \rho V^2] = 0, \quad [\varepsilon + p/\rho + \tfrac{1}{2}V^2] = 0. \tag{50}$$

Solution. On a vortex sheet, $[\boldsymbol{v} \cdot \boldsymbol{n}] = 0$ and we deduce from equations (46), (47) and (48) the discontinuity relations

$$[V] = 0, \quad [\rho] = 0, \quad [\boldsymbol{t}_{(\boldsymbol{n})} \cdot \boldsymbol{n}] = 0, \quad [\boldsymbol{t}_{(\boldsymbol{n})}] + \rho V[\boldsymbol{v}] = \boldsymbol{0}. \tag{A}$$

When the stress is spherical $\boldsymbol{t}_{(\boldsymbol{n})}$ is given by $(18)_1$ and the last two of equations (A) reduce to $[p] = 0$, $\rho V[\boldsymbol{v}] = \boldsymbol{0}$. Since $\rho > 0$ and $[\boldsymbol{v}] \neq \boldsymbol{0}$, V must be zero. The vortex sheet is therefore a material surface and consequently a contact discontinuity.

In a compressible ideal fluid the heat flux vector as well as the stress deviator is identically zero and, with the use of equations $(18)_1$, (36) and (47), we can simplify the jump conditions (48) and (49), obtaining

$$\rho^{\pm}V^{\pm}[\boldsymbol{v}] - [p]\boldsymbol{n} = \boldsymbol{0}, \quad \rho^{\pm}V^{\pm}[\varepsilon + \tfrac{1}{2}\boldsymbol{v}.\boldsymbol{v}] - [p\boldsymbol{v}\,.\,\boldsymbol{n}] = 0. \quad \text{(B)}$$

If the singular surface in question is a wave, V^+ and V^- are both non-zero. Hence, from $(\text{B})_1$, the velocity jump is entirely normal and

$$[p] = \rho^{\pm}V^{\pm}[\boldsymbol{v}\,.\,\boldsymbol{n}] = -\rho^{\pm}V^{\pm}[V], \quad \text{(C)}$$

the last step following from (46). On rewriting the right-hand member of (C) as $-\rho^+V^{+2} + \rho^-V^{-2}$ we arrive at equation $(50)_2$.

We now have

$$[\boldsymbol{v}\,.\,\boldsymbol{v}] = [(\boldsymbol{v}\,.\,\boldsymbol{n})^2 + |\boldsymbol{v} \wedge \boldsymbol{n}|^2] = [(\boldsymbol{v}\,.\,\boldsymbol{n})^2] = [V^2 - 2VV_n],$$

using (46), and

$$\begin{aligned}[p\boldsymbol{v}\,.\,\boldsymbol{n}] &= p^+(V_n - V^+) - p^-(V_n - V^-)\\ &= [p]V_n - (p^+/\rho^+)\rho^-V^- + (p^-/\rho^-)\rho^+V^+\\ &= -\rho^{\pm}V^{\pm}[VV_n + p/\rho],\end{aligned}$$

using in turn (46), (47) and (C). Equation $(50)_3$ is derived by entering these results into $(\text{B})_2$ and cancelling the non-zero factor $\rho^{\pm}V^{\pm}$ on the left-hand side.

Problem 15 A long cylinder is filled with a perfect gas and closed at one end by a plane, rigid piston. Initially the gas is everywhere at rest at density ρ_0 and pressure p_0. A shock wave is then formed by advancing the piston with constant speed U. Assuming that the velocity of the gas to the rear of the shock wave is uniform and parallel to the generators of the cylinder and that no heat is supplied to the gas, determine the speed of propagation of the shock wave and the distributions of density and pressure in the gas.

Solution. The situation at a typical instant following the formation

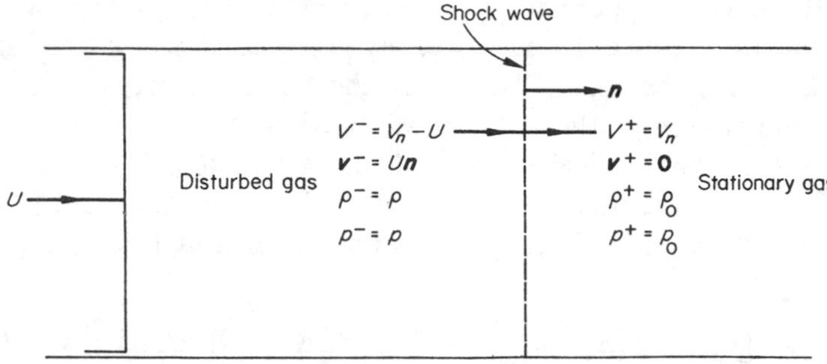

FIGURE 7 The uniform plane shock wave ahead of an advancing piston (Problem 15).

of the shock wave is shown in Figure 7. On the leading face of the piston the speed of the gas must be U and the uniform velocity of the disturbed gas behind the shock wave is therefore given by $\boldsymbol{v}^- = U\boldsymbol{n}$ where $\boldsymbol{n}$ is the unit vector normal to the wave and aligned with the generators of the cylinder as shown. Since the material under consideration is a perfect gas not subject to heat supply the basic jump conditions on the shock wave can be taken in the forms (50) with the internal energy given by equation (40). With the use of (46) we therefore obtain the relations

$$\rho(V_n - U) = \rho_0 V_n, \quad p_0 - p = -\rho_0 V_n U, \tag{A}$$

$$\tfrac{1}{2}V_n^2 + \frac{\gamma}{\gamma - 1}\frac{p_0}{\rho_0} = \tfrac{1}{2}(V_n - U)^2 + \frac{\gamma}{\gamma - 1}\frac{p}{\rho}, \tag{B}$$

in which ρ and p are the uniform values of the density and pressure to the rear of the shock wave and V_n, the local speed of propagation, is also the speed with which the shock wave is entering the stationary gas in front of it.

Set $\alpha = V_n/U$ and $M_0 = U/c_0$ where $c_0 = (\gamma p_0/\rho_0)^{1/2}$. Equations (A) then yield the expressions

$$\rho/\rho_0 = 1 + (\alpha - 1)^{-1}, \quad p/p_0 = 1 + \alpha\gamma M_0^2, \tag{C}$$

from which ρ and p can be calculated when V_n is known. The result of substituting (C) into (B) and simplifying is the quadratic equation

$$2M_0^2\alpha^2 - (\gamma + 1)M_0^2\alpha - 2 = 0,$$

the single positive root of which is

$$\alpha = \tfrac{1}{4}(\gamma + 1) + \{\tfrac{1}{16}(\gamma + 1)^2 + M_0^{-2}\}^{1/2}. \qquad \text{(D)}$$

The speed of propagation of the shock wave is thus given by

$$V_n = \tfrac{1}{4}(\gamma + 1)U + \{\tfrac{1}{16}(\gamma + 1)^2 U^2 + c_0^2\}^{1/2}$$

and the density and pressure of the disturbed gas by equations (C) and (D).

[Certain features of the above results deserve comment. As might have been anticipated, V_n, ρ and p each increase monotonically as the piston speed U increases, but ρ/ρ_0 approaches the finite limit $(\gamma + 1)/(\gamma - 1)$ as $M_0 \to \infty$, indicating the existence of an upper bound to the compression which can be achieved in a perfect gas by passing through it a uniform plane shock wave. As $M_0 \to 0$ we observe that $V_n \to c_0$, $\rho \to \rho_0$ and $p \to p_0$. When $U \ll c_0$ the shock wave is therefore weak in the sense that $[\rho] \ll \rho_0$, $[p] \ll p_0$. The speed c_0 to which V_n then approximates is the *speed of sound* in the undisturbed gas, and the dimensionless constant M_0 expressing U as a multiple of c_0 is hence called the *Mach number* of the piston relative to the initial state of the gas. Finally, to the reader who has noticed that no use has been made in this problem of the pressure–density relation $p = k\rho^\gamma$, we point out that the proportionality factor k takes different values on the two sides of the shock wave. This is associated with a jump in entropy, but we pursue the matter no further since an adequate treatment of continuum thermodynamics lies outside the scope of the present work.]

EXERCISES

(The starred exercises contain results which are used in Chapter 4.)

1. Prove Archimedes' theorem: if a material body is permanently at rest in a configuration B under the action of a uniform gravitational field the resultant contact force on the surface ∂R of an arbitrary subregion R of B is equal in norm and opposite in direction to the weight of the material occupying R.

2*. The stress at a point x is said to be *uniaxial* if the stress vector $\boldsymbol{t}_{(\boldsymbol{n})}$ has a fixed direction independent of $\boldsymbol{n}$. If this fixed direction is defined by the unit vector $\boldsymbol{q}$, show that the stress tensor at x is given by $\boldsymbol{\sigma} = \sigma \boldsymbol{q} \otimes \boldsymbol{q}$ where σ is a scalar. Show also that the normal stress on every surface segment through x is tensile if $\sigma > 0$ and compressive if $\sigma < 0$ and that the maximum shear stress norm at x is $\frac{1}{2}|\sigma|$.

3. Relative to an orthonormal basis e the components of the stress tensor $\boldsymbol{\sigma}$ at a point x are given by

$$\sigma_{ii} = \alpha, \quad \sigma_{ij} = \beta(\neq 0) \quad (i \neq j).$$

Determine the principal stresses, the principal axes of stress and the maximum shear stress norm at x, and show that $-2 < \alpha/\beta \leqslant 1$ is a necessary and sufficient condition for there to be surface segments through x on which the normal stress vanishes.

4. At a point x in the current configuration of a material body the norm of the shear stress on a surface segment whose outward normal is equally inclined to the principal axes of stress is called the *octahedral shear stress* and is denoted by τ_{oct}. Show that

$$\tau_{\text{oct}} = (-\tfrac{2}{3}II_{\boldsymbol{\sigma}'})^{1/2} = \tfrac{1}{3}\{(\sigma_2 - \sigma_3)^2 + (\sigma_3 - \sigma_1)^2 + (\sigma_1 - \sigma_2)^2\}^{1/2},$$

where $\boldsymbol{\sigma}'$ is the stress deviator (see Exercise 1.1, p. 47) and $\sigma_1, \sigma_2, \sigma_3$ are the principal stresses at x.

5. Two symmetric tensors are said to be *coaxial* if their principal axes coincide (i.e. if they have an orthonormal set of proper vectors in common). Prove that the second Piola–Kirchhoff stress $\boldsymbol{t}$ and the right Cauchy–Green strain tensor $\boldsymbol{C}$ are coaxial if and only if the Cauchy stress $\boldsymbol{\sigma}$ is coaxial with the left Cauchy–Green strain tensor $\boldsymbol{B}$.

6. Formulate the balance of angular momentum for a material body acted on by a body torque $\boldsymbol{c}$ per unit mass in addition to the body force $\boldsymbol{b}$, and a contact torque $\boldsymbol{u}_{(\boldsymbol{n})}$ per unit area in addition to the contact force $\boldsymbol{t}_{(\boldsymbol{n})}$. Establish, by a 'tetrahedron argument', the existence of a *couple stress tensor* $\boldsymbol{\mu}$ such that $\boldsymbol{u}_{(\boldsymbol{n})} = \boldsymbol{\mu}^{\mathrm{T}}\boldsymbol{n}$, and go on to derive the field equation and the jump condition associated with the balance law.

7*. A body force is said to be *conservative* if it is expressible as the gradient of a scalar field (called the *body force potential*).

(i) Prove *Kelvin's theorem*: every motion under conservative body forces of a material body composed of an incompressible ideal fluid is circulation preserving, and the same is true of every barotropic motion under conservative body forces of a material body composed of a compressible ideal fluid.

(ii) For a steady motion under conservative body forces of a material body composed of an incompressible ideal fluid, show that the equation of motion can be put into the form

$$\operatorname{grad}(p/\rho + \tfrac{1}{2}\boldsymbol{v}\cdot\boldsymbol{v} - \chi) = -\boldsymbol{\omega} \wedge \boldsymbol{v},$$

where $\boldsymbol{\omega}$ is the vorticity and χ the body force potential. Deduce that $p/\rho + \frac{1}{2}\boldsymbol{v}\cdot\boldsymbol{v} - \chi$ is constant on a streamline [*Bernoulli's equation* for steady motion].

8. For a rigid motion of a material body obtain the expression

$$K(R_t) = \tfrac{1}{2}m(R_t)\bar{\boldsymbol{v}}\cdot\bar{\boldsymbol{v}} + \tfrac{1}{2}\boldsymbol{w}\cdot\{\bar{J}(R_t)\boldsymbol{w}\}$$

for the kinetic energy of the material occupying R_t, the notation being the same as in Problem 3 (p. 91).

9. Show that, in an isochoric motion of a material body $\mathscr{B}$, the kinetic energy of the material occupying an arbitrary regular region R_t can be expressed in the form

$$K(R_t) = \int_{R_t} \rho[\boldsymbol{x}, \boldsymbol{\omega}, \boldsymbol{v}]\,dv + \int_{\partial R_t} \rho\{\tfrac{1}{2}(\boldsymbol{v}\cdot\boldsymbol{v})\boldsymbol{x} - (\boldsymbol{x}\cdot\boldsymbol{v})\boldsymbol{v}\}\cdot\boldsymbol{n}\,da.$$

Deduce that if the motion is also irrotational, the body occupies the same region of $\mathfrak{E}$ at all times and $\boldsymbol{v} = \boldsymbol{0}$ on the boundary of this region, then $\mathscr{B}$ is in a state of rest.

10. Deduce from the basic jump conditions (47) to (49) the following results.

(i) $V^+V^-[\rho] = -[\boldsymbol{t}_{(\boldsymbol{n})}\cdot\boldsymbol{n}]$ [the *Rankine–Hugoniot relation*].
(ii) On a contact discontinuity, $[\boldsymbol{t}_{(\boldsymbol{n})}] = \boldsymbol{0}$ and $[\boldsymbol{v}\cdot\boldsymbol{n}] = 0$. If, in addition, the stress is spherical, $[h_{(\boldsymbol{n})}] = 0$.
(iii) On an acceleration wave, $[\rho] = 0$, $[\boldsymbol{t}_{(\boldsymbol{n})}] = \boldsymbol{0}$ and $\rho V[\varepsilon] + [h_{(\boldsymbol{n})}] = 0$.
(iv) In an isochoric motion no shock wave can exist.

Chapter 4

CONSTITUTIVE EQUATIONS

The balance laws of continuum mechanics, as developed in Chapter 3, make no reference to the constitution of the body concerned beyond presupposing the inability of the material to support contact torques. The specification of the mechanical and thermal properties of a particular material therefore calls for an extension of the theoretical structure presently available to us and it is at this stage that the major subdivisions of the subject, such as the theories of viscous flow and elasticity, branch out from the foundations laid in the last two chapters. The characterization of particular materials is brought within the framework of continuum mechanics through the formulation of constitutive equations (or equations of state). Mathematically the purpose of these relations is to supply connections between kinematic, mechanical and thermal fields which are compatible with the field equations and which, in conjunction with them, yield a theory capable of providing solutions to correctly set problems. Physically, constitutive equations represent various forms of idealized material response which serve as models of the behaviour of actual substances. The predictive value of such models, as assessed experimentally over particular ranges of physical conditions, affords justification for the special continuum theories mentioned above. In the first six sections of this chapter general ideas and procedures underlying the formation of constitutive equations are considered in association with three common examples of idealized mechanical response: ideal fluidity, viscosity and elasticity. Sections 7 and 8 are devoted to problems illustrating the behaviour of incompressible Newtonian viscous fluids and isotropic elastic materials.

1 BASIC CONSTITUTIVE STATEMENT

In relation to an arbitrary basis of E the field equations (3.3), (3.23),

(3.24) and (3.39) provide *eight* scalar relations between *seventeen* scalar fields, namely ρ, ε and the components of $\boldsymbol{v}$, $\boldsymbol{\sigma}$ and $\boldsymbol{q}$, the body force $\boldsymbol{b}$ and the heat supply r being regarded as known. In general, therefore, the field equations do not by themselves determine the behaviour of a material body and this is hardly surprising since given forces and heating are observed to produce a diversity of effects when applied to different substances.

Suppose that a motion of a material body $\mathscr{B}$ is described by equations (2.3) and (2.4), that the density in some reference configuration is prescribed, and that we are provided with recipes for calculating the stress $\boldsymbol{\sigma}$, the internal energy ε and the heat flux $\boldsymbol{q}$, the former as a symmetric tensor function. Then the velocity and the density can be found from equations $(2.12)_1$ and (3.6), and we can go on to compute the terms $\rho\boldsymbol{a}$, div $\boldsymbol{\sigma}$, $\rho\dot{\varepsilon}$, $\mathrm{tr}(\boldsymbol{\sigma D})$ and div $\boldsymbol{q}$ appearing in equations (3.23) and (3.39). The spatial continuity equation (3.3) and the symmetry condition (3.24) are already satisfied, and by appropriately choosing the body force $\boldsymbol{b}$ and the heat supply r we can ensure that the equation of motion and the energy equation also hold. This argument suggests that a natural way of supplying the further relations which must be adjoined to the field equations in order to obtain a complete system of governing equations is to specify $\boldsymbol{\sigma}$, ε and $\boldsymbol{q}$. But if the prescription of these fields were to require a knowledge of no more than the motion of $\mathscr{B}$ the field equations would eventually furnish *five* scalar relations for only *four* scalar fields (ρ and the components of $\boldsymbol{v}$). This prompts the conclusion, again foreseeable, that $\boldsymbol{\sigma}$, ε and $\boldsymbol{q}$ are not, in general, fully determined by the motion of $\mathscr{B}$ but depend also upon a scalar field which is not present as a dependent variable in the field equations. Since ε and $\boldsymbol{q}$ are directly associated with the thermal condition of the body the additional scalar field is expected to measure the heat content or degree of hotness of the material.

With this motivation the *temperature* θ is introduced as a scalar field defined on the configurations of $\mathscr{B}$ and the following *basic constitutive statement* propounded. *At the point* x *occupied at time* t *by an arbitrary particle of* $\mathscr{B}$ *the stress, the internal energy and the heat flux are uniquely determined* (*in conformity with* (3.24)) *by the motion of* $\mathscr{B}$ *together with the temperature on the configurations of* $\mathscr{B}$. The set of *ten* scalar equations implicit in this statement, in conjunction with the *eight* scalar relations provided by the field equations, connect *eighteen* scalar fields, namely ρ, ε, θ and the components of

$\boldsymbol{v}$, $\boldsymbol{\sigma}$ and $\boldsymbol{q}$. The additional equations effectively delimit the response of $\mathscr{B}$ to given applied forces and heating: they thereby specify the constitution of the material composing $\mathscr{B}$ and are aptly referred to as *constitutive equations.*

The basic constitutive statement can be put into mathematical form in a great variety of ways and certain rules have been laid down which admissible constitutive equations must meet. The most important of them are the principles[1] of determinism, local action and objectivity.

The *principle of determinism* states that the values of $\boldsymbol{\sigma}$, ε and $\boldsymbol{q}$ at the point x at time t are uniquely determined by the motion and the temperature field at times up to and including t. The *principle of local action* may be regarded as excluding from constitutive equations effects associated with action at a distance. It asserts that the motions and temperatures experienced by particles which, at time t, occupy points outside an arbitrary neighbourhood of x have no influence on the values at (x, t) of $\boldsymbol{\sigma}$, ε and $\boldsymbol{q}$. According to the *principle of objectivity* the response of a material is the same to any pair of equivalent observers. In mathematical terms material response is described by constitutive equations and an observer is identified with a frame (as defined in Section 3.2). The principle of objectivity therefore requires constitutive equations to be invariant under changes of frame which preserve the essential structure of space and time.

Problem 1 A material body which is able to execute only rigid motions is said to be composed of a *rigid material.* If, for such a material, the internal energy ε is a function of the temperature θ only and the heat flux vector is given by $\boldsymbol{q} = k(\theta)\,\mathrm{grad}\,\theta$, show that the energy equation takes the form

$$\rho c \frac{\partial \theta}{\partial t} = \mathrm{Div}\,(k\,\mathrm{Grad}\,\theta) + \rho r, \qquad \text{(A)}$$

where ρ is the density and $c = \mathrm{d}\varepsilon/\mathrm{d}\theta$.

[1] Although the term *principle* is customarily used in this connection, only one of the precepts mentioned, the principle of determinism, is generally accepted as being consistent with the observed behaviour of all known materials. The status of the other principles is still controversial, but it is not in dispute that representations of idealized response obtained with their aid accurately model the behaviour of extensive classes of fluids and solid materials encountered in practice.

Solution. In a rigid motion the stretching tensor is identically zero (see Problem 2.13, p. 77) and the stress power vanishes. The energy equation (3.39) therefore reduces to

$$\rho c\dot{\theta} = \operatorname{div}(k \operatorname{grad} \theta) + \rho r,$$

use being made also of the given expressions for ε and $\boldsymbol{q}$. When this result is transferred to the referential description $\dot{\theta}$ is replaced by $\partial\theta/\partial t$ and equations $(2.18)_1$ and $(2.19)_1$, with $(2.24)_1$, yield

$$\operatorname{div}(k \operatorname{grad} \theta) = J^{-1} \operatorname{Div}(kJ\boldsymbol{C}^{-1} \operatorname{Grad} \theta).$$

But, in a rigid motion, $\boldsymbol{C} \equiv \boldsymbol{I}$ and $J \equiv 1$ since no stretch and no volume changes occur. Equation (A) is therefore obtained.

[A rigid material provides a further instance of the decoupling of mechanical and thermal effects mentioned in Section 3.5. In this problem constitutive equations for ε and $\boldsymbol{q}$ are supplied and the resulting non-linear partial differential equation for θ is the general form of the *heat conduction equation* for a rigid material, k and c being respectively the *thermal conductivity* and the *specific heat* of the material in question. In the classical theory of heat conduction the temperature dependence of k and c is ignored. Equation (A) then gives way to the linear diffusion equation

$$\rho c \frac{\partial \theta}{\partial t} = k\Delta\theta + \rho r$$

in which $\Delta(\cdot) = \operatorname{Div} \operatorname{Grad}(\cdot)$ is the scalar Laplacian operator in the referential description.]

In the remainder of this chapter three simple, but crucially important, examples of constitutive equations are considered in the context of pure mechanics, that is in the absence of thermo-mechanical interaction. It is assumed in this simplified approach that the temperature is either constant or determined by the motion and that the energy equation is satisfied, equations (3.3) and (3.23) ultimately providing four scalar relations between ρ and the components of $\boldsymbol{v}$. These conditions are fulfilled either exactly or to a close approximation in a wide range of situations to which particular continuum theories, including those mentioned at the beginning of this chapter, have been successfully applied, but a proper examination of their validity involves thermodynamic considerations outside the compass of this book. The restricted form of the basic constitutive statement which we adopt is as follows.

The stress at the point x at time t is uniquely determined (consistently with (3.24)) *by fields derived from the motion of* $\mathscr{B}$ *and also evaluated at* (x, t). (1)

We see that this statement is in accord with the principles of determinism and local action and that, in addition, dependence of the stress upon past details of the motion is precluded. The materials to be considered therefore have no recollection of configurations occupied at former times. For many common fluids and solids, and in particular for air and water over a wide range of conditions and for metals and other structural solids when subjected to moderate forces, this assumption leads to results in good agreement with experiment. But for certain substances, such as polymer solutions, solid plastics and concrete, the exclusion of memory effects can be an over-simplification.

2 EXAMPLES OF CONSTITUTIVE EQUATIONS

The constitutive equation

$$\boldsymbol{\sigma} = -p(\rho)\boldsymbol{I}, \tag{2}$$

already introduced in Section 3.5, characterizes a compressible ideal fluid. This equation clearly conforms to the statement (1) and would continue to do so if the pressure p were allowed to depend upon the referential position $\boldsymbol{X}$ in addition to the density. The functional relationship between p and ρ would then vary from one particle to another and the fluid would be *inhomogeneous.* In the introductory treatment of constitutive equations given in this chapter we restrict attention throughout to homogeneous materials.

The fundamental physical property of a fluid is its inability, when in equilibrium, to sustain shear stresses: fluids obeying the constitutive equation (2) are 'ideal' in the sense that they retain this property when in motion. In the case of a non-ideal fluid it is natural to express the stress in the form $\boldsymbol{\sigma} = -p\boldsymbol{I} + \boldsymbol{\sigma}^{\mathrm{E}}$ where the *extra stress* $\boldsymbol{\sigma}^{\mathrm{E}}$ is zero in a state of rest. In the constitutive equation characterizing a compressible *viscous fluid* p is taken to be a function of the density ρ and $\boldsymbol{\sigma}^{\mathrm{E}}$ is assumed to depend upon ρ and the velocity gradient $\boldsymbol{L}$. Thus

$$\boldsymbol{\sigma} = -\, p(\rho)\boldsymbol{I} + \boldsymbol{f}(\rho, \boldsymbol{L}), \tag{3}$$

where f is a symmetric tensor-valued function. In an ideal fluid the absence of shear stress means that adjacent layers in a simple shearing motion (see Problem 2.12, p. 76) can slip over one another without resistance. In a viscous fluid relative motion of this kind encounters a frictional resistance which Stokes was the first to associate with the rate at which material elements of the fluid are distorted. As shown in Section 2.5, the rate at which local changes of size and shape occur in a motion is specified by the velocity gradient $\boldsymbol{L}$. Since the effects of viscosity are confined to the extra stress, the term ideal fluid used hitherto can be appropriately replaced by *inviscid fluid.*

The constitutive equation which characterizes an *elastic material* is

$$\boldsymbol{\sigma} = \boldsymbol{g}(\boldsymbol{F}), \tag{4}$$

where $\boldsymbol{g}$ is a symmetric tensor-valued function. In contrast to the constitutive equations (2) and (3), each of which connects the stress to the local motion independently of any reference configuration, equation (4), through the appearance of the deformation gradient, involves the choice of some such configuration. Since the choice is arbitrary, (4) may be interpreted thus: at a representative particle of an elastic material changes of stress take place in response to changes of configuration regardless of the intermediate states occupied by the material in passing from one configuration to another. Alternatively, an elastic material may be particularized as one whose stress response is, in principle, deducible from statical experiments involving only homogeneous deformations (see Section 2.4).

In common with (2), equations (3) and (4) are consistent with the basic statement (1), but there remains the question of their compatibility with the principle of objectivity. We return to this point in Section 4.

3 OBSERVER TRANSFORMATIONS

It has already been noted that an observer is characterized mathematically as a frame and is therefore equipped to measure position in the Euclidean point space $\mathfrak{E}$ and time on the real line R. Two observers are said to be equivalent if they agree about (*a*) the distance between an arbitrary pair of points in $\mathfrak{E}$, (*b*) orientation (i.e. right- or left-handedness) in $\mathfrak{E}$, (*c*) the time elapsed between an arbitrary pair of instants in R, and (*d*) the order in which two distinct instants occur.

Let $\mathscr{F}$ denote a frame in which position, relative to an origin o, and time are $\boldsymbol{x}$ and t respectively, and let $\mathscr{F}'$ be a second frame in which position, relative to an origin o', is $\boldsymbol{x}'$ and time t'. These two frames satisfy the invariance conditions (*a*) to (*d*) above if and only if $(\boldsymbol{x}, t)$ and $(\boldsymbol{x}', t')$ are connected by the relations

$$\boldsymbol{x}' = \boldsymbol{c}(t) + \boldsymbol{Q}(t)\boldsymbol{x}, \quad t' = t - a, \tag{5}$$

where $\boldsymbol{Q}$ is a proper orthogonal tensor and a is a constant. Equations (5) define a mapping of space–time (i.e. $\mathfrak{E} \times \mathsf{R}$) into itself which we refer to as the *observer transformation* $\mathscr{F} \to \mathscr{F}'$. Evidently an observer transformation effects a change of the spatial description.

Equation $(5)_1$ can be deduced from the spatial requirements (*a*) and (*b*)[2] by a straightforward adaptation of the solution to the first part of Problem 2.1 (p. 52), while $(5)_2$ is a trivial consequence of (*c*) and (*d*). The reader is recommended to write out proofs and also to verify that (5) are sufficient conditions for the four invariance properties to hold.

On solving equations (5) for $\boldsymbol{x}$ and t we obtain

$$\boldsymbol{x} = \boldsymbol{Q}^{\mathrm{T}}(t' + a)\{\boldsymbol{x}' - \boldsymbol{c}(t' + a)\}, \quad t = t' + a. \tag{6}$$

The inverse mapping $\mathscr{F}' \to \mathscr{F}$ defined by these relations is also an observer transformation and it is easily shown that if $\mathscr{F} \to \mathscr{F}'$ and $\mathscr{F}' \to \mathscr{F}''$ are observer transformations, then so is $\mathscr{F} \to \mathscr{F}''$. Thus equivalence of observers, as defined at the beginning of this section, is indeed an equivalence relation, and we denote by $\mathscr{E}(\mathscr{F})$ the set of all frames obtainable from a given frame $\mathscr{F}$ by observer transformations. Henceforth $\mathscr{F}'$ signifies an arbitrary member of $\mathscr{E}(\mathscr{F})$ and it is understood that the observer transformation $\mathscr{F} \to \mathscr{F}'$ has the representation (5).

Suppose now that a scalar field, defined on some domain, is denoted by ϕ when referred to $\mathscr{F}$ and ϕ' when referred to $\mathscr{F}'$ and

[2] If the orientation condition (*b*) is omitted, equation $(5)_1$ remains valid, but the orthogonal tensor $\boldsymbol{Q}$ need no longer be proper. In fact none of the results given subsequently are affected by this change.

that $\boldsymbol{u}$, $\boldsymbol{u}'$ and $\boldsymbol{T}$, $\boldsymbol{T}'$ have analogous meanings in relation to a vector and a tensor field on the same domain. If, for all $\mathscr{F}' \in \mathscr{E}(\mathscr{F})$,

$$\left.\begin{aligned} \phi'(\boldsymbol{x}', t') = \phi(\boldsymbol{x}, t), \quad \boldsymbol{u}'(\boldsymbol{x}', t') = \boldsymbol{Q}(t)\boldsymbol{u}(\boldsymbol{x}, t), \\ \boldsymbol{T}'(\boldsymbol{x}', t') = \boldsymbol{Q}(t)\boldsymbol{T}(\boldsymbol{x}, t)\boldsymbol{Q}^{\mathrm{T}}(t), \end{aligned}\right\} \tag{7}$$

the three fields are said to be *objective*. The transformation rules $(7)_{2,3}$ ensure that the directions associated with the vector $\boldsymbol{u}$ and the tensor $\boldsymbol{T}$ are unaltered by an observer transformation. To see this we note that the triad of directions defined in $\mathscr{F}$ by the orthonormal basis $e = \{\boldsymbol{e}_1, \boldsymbol{e}_2, \boldsymbol{e}_3\}$ coincide with those defined in $\mathscr{F}'$ by the similar basis $e' = \{\boldsymbol{e}'_1, \boldsymbol{e}'_2, \boldsymbol{e}'_3\}$ where

$$\boldsymbol{e}'_i(t) = \boldsymbol{Q}(t)\boldsymbol{e}_i. \tag{8}$$

Thus $\boldsymbol{Q}(t) = \boldsymbol{e}'_p(t) \otimes \boldsymbol{e}_p$ (cf. Exercise 1.10, p. 48), and on substituting this expression into equations $(7)_{2,3}$ we duly find that the components of the vector and tensor fields relative to e in the frame $\mathscr{F}$ and to e' in $\mathscr{F}'$ are equal:

$$\boldsymbol{u}'(\boldsymbol{x}', t').\boldsymbol{e}'_i = \boldsymbol{u}(\boldsymbol{x}, t).\boldsymbol{e}_i, \quad \boldsymbol{e}'_i.\{\boldsymbol{T}'(\boldsymbol{x}', t')\boldsymbol{e}'_j\} = \boldsymbol{e}_i.\{\boldsymbol{T}(\boldsymbol{x}, t)\boldsymbol{e}_j\}. \tag{9}$$

Problem 2 Prove that

(i) the spatial gradient of an objective scalar field is an objective vector field,

(ii) the spatial divergence of an objective vector field is an objective scalar field, and

(iii) the spatial divergence of an objective tensor field is an objective vector field.

Solution. Let (o, e) and (o', e') be rectangular Cartesian coordinate systems in $\mathscr{F}$ and $\mathscr{F}'$ respectively, the base vectors being connected by (8). Then equation $(5)_1$ yields

$$(\boldsymbol{x}' - \boldsymbol{c}) \cdot \boldsymbol{e}'_i = \boldsymbol{e}'_i \cdot (\boldsymbol{Q}\boldsymbol{x}) = \boldsymbol{x} \cdot (\boldsymbol{Q}^{\mathrm{T}}\boldsymbol{e}'_i) = \boldsymbol{x} \cdot \boldsymbol{e}_i$$

which tells us that $x'_i = \boldsymbol{x}' \cdot \boldsymbol{e}'_i$, the coordinates in the system (o', e') of an arbitrary point, differ from $x_i = \boldsymbol{x} \cdot \boldsymbol{e}_i$, their counterparts in (o, e), by functions of time only. If ϕ, $\boldsymbol{u}$, $\boldsymbol{T}$ and ϕ', $\boldsymbol{u}'$, $\boldsymbol{T}'$ are the forms taken in $\mathscr{F}$ and $\mathscr{F}'$ by the objective scalar, vector and tensor fields under consideration we deduce from equations $(7)_1$ and (9) that

$$\left.\begin{aligned} \frac{\partial \phi'}{\partial x'_i}(\boldsymbol{x}', t') = \frac{\partial \phi}{\partial x_i}(\boldsymbol{x}, t), \quad \frac{\partial u'_i}{\partial x'_j}(\boldsymbol{x}', t') = \frac{\partial u_i}{\partial x_j}(\boldsymbol{x}, t), \\ \frac{\partial T'_{ij}}{\partial x'_k}(\boldsymbol{x}', t') = \frac{\partial T_{ij}}{\partial x_k}(\boldsymbol{x}, t), \end{aligned}\right\} \tag{A}$$

the components u_i, T_{ij} relating to e and u'_i, T'_{ij} to e'.

(i) Using equations (1.92), $(A)_1$ and (8), we now have

$$(\text{grad}\, \phi')(\boldsymbol{x}', t') = \frac{\partial \phi'}{\partial x'_p}(\boldsymbol{x}', t')\boldsymbol{e}'_p = \frac{\partial \phi}{\partial x_p}(\boldsymbol{x}, t)\boldsymbol{Q}(t)\boldsymbol{e}_p = \boldsymbol{Q}(t)(\text{grad}\, \phi)(\boldsymbol{x}, t).$$

(ii) Next, from equations $(1.93)_2$ and $(A)_2$,

$$(\text{div}\, \boldsymbol{u}')(\boldsymbol{x}', t') = \frac{\partial u'_p}{\partial x'_p}(\boldsymbol{x}', t') = \frac{\partial u_p}{\partial x_p}(\boldsymbol{x}, t) = (\text{div}\, \boldsymbol{u})(\boldsymbol{x}, t).$$

(iii) And lastly, equations (1.94), $(A)_3$ and (8) give

$$(\text{div}\, \boldsymbol{T}')(\boldsymbol{x}', t') = \frac{\partial T'_{pq}}{\partial x'_p}(\boldsymbol{x}', t')\boldsymbol{e}'_q = \frac{\partial T_{pq}}{\partial x_p}(\boldsymbol{x}, t)\boldsymbol{Q}(t)\boldsymbol{e}_q = \boldsymbol{Q}(t)\,(\text{div}\, \boldsymbol{T})\,(\boldsymbol{x}, t).$$

Problem 3 With reference to a moving body, obtain equations describing the behaviour of the velocity $\boldsymbol{v}$, the acceleration $\boldsymbol{a}$, the deformation gradient $\boldsymbol{F}$ and the velocity gradient $\boldsymbol{L}$ under an observer transformation.

Solution. Suppose that the motion is represented in $\mathscr{F}$ by equation (2.3). Then it is given in $\mathscr{F}'$ by

$$\boldsymbol{x}' = \boldsymbol{c}(t' + a) + \boldsymbol{Q}(t' + a)\boldsymbol{\psi}(\boldsymbol{X}, t' + a) = \boldsymbol{\psi}'(\boldsymbol{X}, t'), \quad \text{say,} \tag{10}$$

use being made of equations (5).

(*a*) Differentiation of (10) with respect to t' yields the following connections between the velocities and accelerations in $\mathscr{F}$ and $\mathscr{F}'$:

$$\boldsymbol{v}'(\boldsymbol{x}', t') = \frac{\partial \boldsymbol{\psi}'}{\partial t'}(\boldsymbol{X}, t')$$

$$= \dot{\boldsymbol{c}}(t' + a) + \dot{\boldsymbol{Q}}(t' + a)\boldsymbol{\psi}(\boldsymbol{X}, t' + a) + \boldsymbol{Q}(t' + a)\frac{\partial \boldsymbol{\psi}}{\partial t}(\boldsymbol{X}, t' + a)$$

$$= \dot{\boldsymbol{c}}(t) + \dot{\boldsymbol{Q}}(t)\boldsymbol{x} + \boldsymbol{Q}(t)\boldsymbol{v}(\boldsymbol{x}, t)$$

and

$$\boldsymbol{a}'(\boldsymbol{x}', t') = \frac{\partial^2 \boldsymbol{\psi}'}{\partial t'^2}(\boldsymbol{X}, t') = \ddot{\boldsymbol{c}}(t) + \ddot{\boldsymbol{Q}}(t)\boldsymbol{x} + 2\dot{\boldsymbol{Q}}(t)\boldsymbol{v}(\boldsymbol{x}, t) + \boldsymbol{Q}(t)\boldsymbol{a}(\boldsymbol{x}, t).$$

It can be shown, as in Problem 2.1 (p. 52), that $\boldsymbol{\Omega} = \dot{\boldsymbol{Q}}\boldsymbol{Q}^{\mathrm{T}}$ is a skew-symmetric tensor (representing the spin of $\mathscr{F}$ relative to $\mathscr{F}'$) and that $\ddot{\boldsymbol{Q}}\boldsymbol{Q}^{\mathrm{T}} = \dot{\boldsymbol{\Omega}} + \boldsymbol{\Omega}^2$. With the aid of these results and equations (6) the transformation formulae for the velocity and acceleration can be expressed in the forms

$$\boldsymbol{v}' = \dot{\boldsymbol{c}} + \boldsymbol{Q}\boldsymbol{v} + \boldsymbol{\Omega}(\boldsymbol{x}' - \boldsymbol{c}), \quad \boldsymbol{a}' = \ddot{\boldsymbol{c}} + \boldsymbol{Q}\boldsymbol{a} + 2\boldsymbol{\Omega}(\boldsymbol{v}' - \dot{\boldsymbol{c}}) + (\dot{\boldsymbol{\Omega}} - \boldsymbol{\Omega}^2)(\boldsymbol{x}' - \boldsymbol{c}). \qquad (11)$$

(*b*) The result of applying the referential gradient operator to (10) is

$$\boldsymbol{F}(\boldsymbol{X}, t') = (\mathrm{Grad}\, \boldsymbol{\psi}')(\boldsymbol{X}, t') = \boldsymbol{Q}(t' + a)(\mathrm{Grad}\, \boldsymbol{\psi})(\boldsymbol{X}, t' + a)$$

$$= \boldsymbol{Q}(t)(\mathrm{Grad}\, \boldsymbol{\psi})(\boldsymbol{X}, t) = \boldsymbol{Q}(t)\boldsymbol{F}(\boldsymbol{X}, t).$$

The deformation gradients in $\mathscr{F}$ and $\mathscr{F}'$ are therefore related by

$$\boldsymbol{F}' = \boldsymbol{Q}\boldsymbol{F}. \qquad (12)$$

(*c*) On differentiating equation (12) with respect to t' we obtain

$$\dot{\boldsymbol{F}}'(\boldsymbol{X}, t') = \boldsymbol{Q}(t)\dot{\boldsymbol{F}}(\boldsymbol{X}, t) + \dot{\boldsymbol{Q}}(t)\boldsymbol{F}(\boldsymbol{X}, t).$$

Hence, using equations $(2.30)_2$ and (12) and suppressing the arguments of functions,

$$\boldsymbol{L}' = \dot{\boldsymbol{F}}'\boldsymbol{F}'^{-1} = (\boldsymbol{Q}\dot{\boldsymbol{F}} + \dot{\boldsymbol{Q}}\boldsymbol{F})(\boldsymbol{Q}\boldsymbol{F})^{-1} = \boldsymbol{Q}\dot{\boldsymbol{F}}\boldsymbol{F}^{-1}\boldsymbol{Q}^{\mathrm{T}} + \dot{\boldsymbol{Q}}\boldsymbol{Q}^{\mathrm{T}}$$

$$= \boldsymbol{Q}\boldsymbol{L}\boldsymbol{Q}^{\mathrm{T}} + \boldsymbol{\Omega}. \qquad (13)$$

This is the required transformation formula for the velocity gradient.

Inspection of equations (11) to (13) shows that each of the fields examined in Problem 3 fails to be objective. However, it follows from equations (12) and $(2.24)_2$ that the behaviour of the left Cauchy–Green strain tensor under an observer transformation is given by

$$\boldsymbol{B}' = \boldsymbol{F}'\boldsymbol{F}'^{\mathrm{T}} = (\boldsymbol{QF})(\boldsymbol{QF})^{\mathrm{T}} = \boldsymbol{QFF}^{\mathrm{T}}\boldsymbol{Q}^{\mathrm{T}} = \boldsymbol{QBQ}^{\mathrm{T}},$$

while (13) and $(2.43)_2$ supply the transformation formula

$$\boldsymbol{D}' = \tfrac{1}{2}(\boldsymbol{L}' + \boldsymbol{L}'^{\mathrm{T}}) = \tfrac{1}{2}(\boldsymbol{QLQ}^{\mathrm{T}} + \boldsymbol{QL}^{\mathrm{T}}\boldsymbol{Q}^{\mathrm{T}} + \boldsymbol{\Omega} + \boldsymbol{\Omega}^{\mathrm{T}}) = \boldsymbol{QDQ}^{\mathrm{T}}$$

for the stretching tensor. Thus $\boldsymbol{B}$ and $\boldsymbol{D}$ are objective tensors, a conclusion of some moment in relation to the later discussion of constitutive equations.

Problem 4 ϕ is a scalar field, $\boldsymbol{u}$ a vector field and $\boldsymbol{T}$ a tensor field, each defined on the configurations of a moving body $\mathscr{B}$. The *convected rates of change* of $\boldsymbol{u}$ and $\boldsymbol{T}$ are defined by

$$\hat{\boldsymbol{u}} = \dot{\boldsymbol{u}} + \boldsymbol{L}^{\mathrm{T}}\boldsymbol{u}, \quad \mathring{\boldsymbol{T}} = \dot{\boldsymbol{T}} + \boldsymbol{L}^{\mathrm{T}}\boldsymbol{T} + \boldsymbol{TL}. \tag{14}$$

If ϕ, $\boldsymbol{u}$ and $\boldsymbol{T}$ are objective, prove that the same is true of $\dot{\phi}$, $\hat{\boldsymbol{u}}$ and $\mathring{\boldsymbol{T}}$.

Solution. (*a*) An objective scalar field defined on the configurations of $\mathscr{B}$ satisfies equation $(7)_1$ which, via (2.3) and (10), can be restated in the referential form

$$\phi'(\psi'(X, t'), t') = \phi(\psi(X, t), t).$$

Differentiating each side of this relation with respect to t' then reverting to the spatial descriptions in the frames $\mathscr{F}$ and $\mathscr{F}'$, we obtain, with the use of equations $(2.10)_1$ and $(6)_2$,

$$\dot{\phi}'(\boldsymbol{x}', t') = \dot{\phi}(\boldsymbol{x}, t)\,(\partial t/\partial t') = \dot{\phi}(\boldsymbol{x}, t).$$

Thus $\dot{\phi}$ is an objective scalar.

(*b*) Suppose that, as in (*a*), equations (2.3) and (10) have been used to express $\boldsymbol{u}$, $\boldsymbol{T}$ and $\boldsymbol{u}'$, $\boldsymbol{T}'$ as functions of the referential position $\boldsymbol{X}$ and either t or t'. Then, omitting the arguments of functions,

$$\hat{\boldsymbol{u}}' = (\partial \boldsymbol{u}'/\partial t') + \boldsymbol{L}'^{\mathrm{T}}\boldsymbol{u}' = (\partial/\partial t)(\boldsymbol{Qu}) + (\boldsymbol{QL}^{\mathrm{T}}\boldsymbol{Q}^{\mathrm{T}} - \boldsymbol{\Omega})(\boldsymbol{Qu})$$

$$= \boldsymbol{Q}(\dot{\boldsymbol{u}} + \boldsymbol{L}^{\mathrm{T}}\boldsymbol{u}) + (\dot{\boldsymbol{Q}}\boldsymbol{Q}^{\mathrm{T}} - \boldsymbol{\Omega})\boldsymbol{Qu} = \boldsymbol{Q}\hat{\boldsymbol{u}},$$

use being made here of equations $(6)_2$, $(7)_2$, (13), $(14)_1$ and the relation $\boldsymbol{\Omega} = \dot{\boldsymbol{Q}}\boldsymbol{Q}^{\mathrm{T}}$. Similarly, employing equations $(6)_2$, $(7)_3$, (13) and $(14)_2$, we find that

$$\begin{aligned}\mathring{\boldsymbol{T}}' &= (\partial \boldsymbol{T}'/\partial t') + \boldsymbol{L}'^{\mathrm{T}}\boldsymbol{T}' + \boldsymbol{T}'\boldsymbol{L}' \\ &= (\partial/\partial t)(\boldsymbol{Q}\boldsymbol{T}\boldsymbol{Q}^{\mathrm{T}}) + (\boldsymbol{Q}\boldsymbol{L}^{\mathrm{T}}\boldsymbol{Q}^{\mathrm{T}} - \boldsymbol{\Omega})\boldsymbol{Q}\boldsymbol{T}\boldsymbol{Q}^{\mathrm{T}} + \boldsymbol{Q}\boldsymbol{T}\boldsymbol{Q}^{\mathrm{T}}(\boldsymbol{Q}\boldsymbol{L}\boldsymbol{Q}^{\mathrm{T}} + \boldsymbol{\Omega}) \\ &= \boldsymbol{Q}(\dot{\boldsymbol{T}} + \boldsymbol{L}^{\mathrm{T}}\boldsymbol{T} + \boldsymbol{T}\boldsymbol{L})\boldsymbol{Q}^{\mathrm{T}} + (\dot{\boldsymbol{Q}}\boldsymbol{Q}^{\mathrm{T}} - \boldsymbol{\Omega})\boldsymbol{Q}\boldsymbol{T}\boldsymbol{Q}^{\mathrm{T}} + \boldsymbol{Q}\boldsymbol{T}\boldsymbol{Q}^{\mathrm{T}}(\boldsymbol{Q}\dot{\boldsymbol{Q}}^{\mathrm{T}} + \boldsymbol{\Omega}) \\ &= \boldsymbol{Q}\mathring{\boldsymbol{T}}\boldsymbol{Q}^{\mathrm{T}}.\end{aligned}$$

The convected rates of change of objective vector and tensor fields are therefore themselves objective.

The results of Problems 2 to 4 can be used to study the effect of an observer transformation on the basic field equations derived in Chapter 3. But first it is necessary to specify the behaviour under such transformations of the fields which represent the primitive concepts of mass, force, internal energy and heating. Of these the density ρ, the stress vector $\boldsymbol{t}_{(\boldsymbol{n})}$, the internal energy ε and the heat flux $h_{(\boldsymbol{n})}$, being concerned with the internal circumstances of a material body, are expected to appear the same to equivalent observers. They are accordingly taken to be objective, an assumption which may properly be viewed as part of the principle of objectivity. It follows that the stress tensor $\boldsymbol{\sigma}$ and the heat flux vector $\boldsymbol{q}$ are objective and hence, from Problems 2 to 4, that each of the terms $\dot{\rho}$, $\rho \operatorname{div} \boldsymbol{v}$ $(= \rho \operatorname{tr} \boldsymbol{D})$, $\operatorname{div}\boldsymbol{\sigma}$, $\rho\dot{\varepsilon}$, $\operatorname{tr}(\boldsymbol{\sigma}\boldsymbol{D})$ and $\operatorname{div} \boldsymbol{q}$ appearing in the field equations is also objective. Thus the spatial continuity equation (3.3) and the symmetry condition (3.24) retain their validity under an arbitrary observer transformation, as does the energy equation (3.39) subject to the further assumption that the heat supply r is an objective scalar field. The situation regarding the equation of motion (3.23) is less straightforward since, from $(11)_2$, the inertial term $\rho\boldsymbol{a}$ is not an objective vector. It is asserted in the statement of the laws of motion (Section 3.2) that there is a frame $\mathscr{F}$ in which this equation holds (it is called an *inertial frame*). Since $\operatorname{div}\boldsymbol{\sigma}$ is objective the validity of (3.23) in a different frame in $\mathscr{E}(\mathscr{F})$ requires that $\boldsymbol{a} - \boldsymbol{b}$ shall transform as an objective vector and this means that the terms additional to $\boldsymbol{Q}\boldsymbol{a}$ on the right-hand side of equation $(11)_2$ become associated with the body force in the new frame. The fictitious forces which arise in this way are known as *inertial forces*.

4 REDUCED CONSTITUTIVE EQUATIONS

In the light of the developments described in Section 3 the principle of objectivity can be restated as follows: constitutive equations formulated in a frame $\mathscr{F}$ are invariant under the class of observer transformations $\{\mathscr{F} \to \mathscr{F}' : \mathscr{F}' \in \mathscr{E}(\mathscr{F})\}$. We now consider the bearing of this requirement on the constitutive equations (2), (3) and (4) introduced in Section 2.

Turning first to equation (2), regarded now as holding in $\mathscr{F}$, we utilize the objectivity of the stress and the density to obtain

$$\boldsymbol{\sigma}' = \boldsymbol{Q}\boldsymbol{\sigma}\boldsymbol{Q}^{\mathrm{T}} = -p(\rho)\boldsymbol{Q}\boldsymbol{I}\boldsymbol{Q}^{\mathrm{T}} = -p(\rho)\boldsymbol{I} = -p(\rho')\boldsymbol{I}. \tag{15}$$

Here primes continue to designate fields referred to a second frame $\mathscr{F}'$, selected arbitrarily from $\mathscr{E}(\mathscr{F})$, and $\boldsymbol{Q}$ is the proper orthogonal tensor associated, through equation $(5)_1$, with the observer transformation $\mathscr{F} \to \mathscr{F}'$. The equalities (15) confirm that if (2) holds in $\mathscr{F}$ then it also holds in $\mathscr{F}'$; that, in other words, the constitutive equation characterizing an inviscid fluid conforms to the principle of objectivity regardless of the functional form in which the pressure depends upon the density.

It is apparent from the transformation formulae (13) and (12) that the constitutive equations (3) and (4) will be objective only if the *response functions* $\boldsymbol{f}$ and $\boldsymbol{g}$ are restricted in some way. The nature of these restrictions is exposed in the following problem.

Problem 5 Prove that the constitutive equations (3) and (4) are compatible with the principle of objectivity if and only if

$$\boldsymbol{f}(\rho, \boldsymbol{Q}\boldsymbol{L}\boldsymbol{Q}^{\mathrm{T}} + \boldsymbol{\Omega}) = \boldsymbol{Q}\boldsymbol{f}(\rho, \boldsymbol{L})\boldsymbol{Q}^{\mathrm{T}} \tag{16}$$

and

$$\boldsymbol{g}(\boldsymbol{Q}\boldsymbol{F}) = \boldsymbol{Q}\boldsymbol{g}(\boldsymbol{F})\boldsymbol{Q}^{\mathrm{T}} \tag{17}$$

for all proper orthogonal tensors $\boldsymbol{Q}$ and (in the case of (16)) for all skew-symmetric tensors $\boldsymbol{\Omega}$.

Solution. (*a*) It has been shown above that the contribution $-p(\rho)\boldsymbol{I}$ to the stress in equation (3) is objective. The extra stress $\boldsymbol{\sigma}^{\mathrm{E}}$ is therefore an objective tensor and it suffices to show that (16) is a necessary and sufficient condition for the constitutive equation

$$\boldsymbol{\sigma}^{\mathrm{E}} = \boldsymbol{f}(\rho, \boldsymbol{L}) \tag{A}$$

to comply with the principle of objectivity.

To demonstrate the necessity of (16), suppose that equation (A) is objective. Then

$$\boldsymbol{\sigma}^{\mathrm{E}} = \boldsymbol{f}(\rho, \boldsymbol{L}) \text{ in } \mathscr{F} \quad \text{and} \quad \boldsymbol{\sigma}'^{\mathrm{E}} = \boldsymbol{f}(\rho', \boldsymbol{L}') \text{ in } \mathscr{F}' \tag{B}$$

for all $\mathscr{F}' \in \mathscr{E}(\mathscr{F})$. In addition, $\boldsymbol{\sigma}'^{\mathrm{E}} = \boldsymbol{Q}\boldsymbol{\sigma}^{\mathrm{E}}\boldsymbol{Q}^{\mathrm{T}}$, $\rho' = \rho$, and the connection (13) applies to $\boldsymbol{L}'$ and $\boldsymbol{L}$. Equation (16) is obtained on combining these relations with (B) and realizing that the frame $\mathscr{F}'$ can be so chosen from $\mathscr{E}(\mathscr{F})$ that the proper orthogonal tensor $\boldsymbol{Q}$ involved in the observer transformation $\mathscr{F} \to \mathscr{F}'$ and its derivative $\dot{\boldsymbol{Q}}(= \boldsymbol{\Omega}\boldsymbol{Q})$ take on any assigned values.

Conversely, if (16) is valid for all proper orthogonal $\boldsymbol{Q}$ and all skew-symmetric $\boldsymbol{\Omega}$ and the constitutive equation (A) holds in some frame $\mathscr{F}$, the extra stress in an arbitrary equivalent frame $\mathscr{F}'$ is

$$\boldsymbol{\sigma}'^{\mathrm{E}} = \boldsymbol{Q}\boldsymbol{\sigma}^{\mathrm{E}}\boldsymbol{Q}^{\mathrm{T}} = \boldsymbol{Q}\boldsymbol{f}(\rho, \boldsymbol{L})\boldsymbol{Q}^{\mathrm{T}} = \boldsymbol{f}(\rho, \boldsymbol{Q}\boldsymbol{L}\boldsymbol{Q}^{\mathrm{T}} + \boldsymbol{\Omega}) = \boldsymbol{f}(\rho', \boldsymbol{L}'),$$

appeal being made to the objectivity of $\boldsymbol{\sigma}^{\mathrm{E}}$ and ρ and to equation (13). Thus (A) holds in $\mathscr{F}'$ and it follows that (16) is a sufficient as well as a necessary condition for equation (3) to be objective.

(*b*) The establishment of (17) as a necessary and sufficient condition for equation (4) to satisfy the principle of objectivity proceeds along the lines followed in (*a*), equation (12) being used in place of (13). The reader is left to supply the details.

At a representative particle of a viscous fluid equation (16) holds for all proper orthogonal $\boldsymbol{Q}$ and all skew-symmetric $\boldsymbol{\Omega}$. The particular choice $\boldsymbol{Q} = \boldsymbol{I}, \boldsymbol{\Omega} = \frac{1}{2}(\boldsymbol{L}^{\mathrm{T}} - \boldsymbol{L})$ leads to the identity

$$\boldsymbol{f}(\rho, \boldsymbol{L}) = \boldsymbol{f}(\rho, \tfrac{1}{2}(\boldsymbol{L} + \boldsymbol{L}^{\mathrm{T}})), \tag{18}$$

and hence to the conclusion that the stress in a viscous fluid depends upon the velocity gradient through the stretching tensor $\boldsymbol{D}$. The constitutive equation (3) can therefore be replaced by

$$\boldsymbol{\sigma} = -p(\rho)\boldsymbol{I} + \boldsymbol{f}(\rho, \boldsymbol{D}) \tag{19}$$

and, in view of (18), the restriction (16) on the response function $\boldsymbol{f}$ becomes

$$\boldsymbol{f}(\rho, \boldsymbol{Q}\boldsymbol{D}\boldsymbol{Q}^{\mathrm{T}}) = \boldsymbol{Q}\boldsymbol{f}(\rho, \boldsymbol{D})\boldsymbol{Q}^{\mathrm{T}} \quad \forall \text{ proper orthogonal } \boldsymbol{Q}. \tag{20}$$

A corresponding reduction of the constitutive equation (4) results from the condition (17). Replacing F by its right polar decomposition RU on the left-hand side of (17) and setting $Q = R^{\mathrm{T}}$ we find that

$$g(F) = Rg(U)R^{\mathrm{T}}. \tag{21}$$

In fact (17), regarded as a functional equation for g, is solved by (21), as can be verified by the direct substitution of the identity

$$g(F) = F(F^{\mathrm{T}}F)^{-1/2}g((F^{\mathrm{T}}F)^{1/2})\,(F^{\mathrm{T}}F)^{-1/2}F^{\mathrm{T}}$$

which is obtained from (21) by writing $R = FU^{-1}$ and recalling that U is the unique positive definite square root of $C = F^{\mathrm{T}}F$. The constitutive equation of an elastic material therefore reduces to

$$\sigma = Rg(U)R^{\mathrm{T}}, \tag{22}$$

and in the transition from (4) to (22) the effects of stretch and rotation on the stress at a typical particle have been separated out. The dependence of σ on the right stretch tensor U is seen to be unrestricted, but the rotation necessarily enters the stress–deformation relation in the explicit manner displayed in equation (22).

5 MATERIAL SYMMETRY

Suppose that a body, composed of elastic material, is subjected to a given deformation in relation to a specified reference configuration. If, with an alternative choice of reference configuration, the same deformation is again applied the stress produced at a representative particle $\mathscr{X}$ will generally be different. However, most elastic materials exhibit *material symmetry* in the sense that particular changes of reference configuration exist which leave the stress at $\mathscr{X}$ arising from an arbitrary deformation invariant. The larger the collection of such transformations the greater the degree of symmetry possessed by the material.

In order to develop the concept of material symmetry mathematically it is first necessary to determine the effect of a change of reference configuration on the deformation gradient. Let B_r and B_r^* be an arbitrary pair of configurations of a body $\mathscr{B}$ and let F and F^* be the deformation gradients relating the current configuration of $\mathscr{B}$ to B_r and B_r^* regarded as reference states. In relation to rectangular

Cartesian systems (O, E), (O^*, E^*) and (o, e) of referential and spatial coordinates, $\boldsymbol{F}$ and $\boldsymbol{F}^*$ are given by

$$\boldsymbol{F} = \frac{\partial x_p}{\partial X_\pi} \boldsymbol{e}_p \otimes \boldsymbol{E}_\pi, \quad \boldsymbol{F}^* = \frac{\partial x_p}{\partial X^*_\pi} \boldsymbol{e}_p \otimes \boldsymbol{E}^*_\pi$$

(cf. equation (2.14)). Hence, using the chain rule for partial differentiation and the orthonormality of the basis $\boldsymbol{E}^*$, we obtain

$$\boldsymbol{F} = \frac{\partial x_p}{\partial X^*_\rho} \frac{\partial X^*_\rho}{\partial X_\pi} \boldsymbol{e}_p \otimes \boldsymbol{E}_\pi = \left(\frac{\partial x_p}{\partial X^*_\rho} \boldsymbol{e}_p \otimes \boldsymbol{E}^*_\rho\right)\left(\frac{\partial X^*_\sigma}{\partial X_\pi} \boldsymbol{E}^*_\sigma \otimes \boldsymbol{E}_\pi\right) = \boldsymbol{F}^*\boldsymbol{P}, \tag{23}$$

$\boldsymbol{P}$ being the deformation gradient associated with the mapping $B_r \to B^*_r$.

Implicit in the response function $\boldsymbol{g}$ in the constitutive equation (4) is the reference configuration B_r relative to which the deformation gradient $\boldsymbol{F}$ is defined. With the alternative choice B^*_r equation (4) takes the form

$$\boldsymbol{\sigma} = \boldsymbol{g}^*(\boldsymbol{F}^*), \tag{24}$$

where, in view of the relation (23), the response function $\boldsymbol{g}^*$ is defined by

$$\boldsymbol{g}^*(\boldsymbol{A}) = \boldsymbol{g}(\boldsymbol{A}\boldsymbol{P}) \quad \forall \text{ invertible } \boldsymbol{A}. \tag{25}$$

Suppose now that there is an invertible tensor $\boldsymbol{P}$ such that

$$\boldsymbol{g}(\boldsymbol{A}\boldsymbol{P}) = \boldsymbol{g}(\boldsymbol{A}) \quad \forall \text{ invertible } \boldsymbol{A}. \tag{26}$$

Then, interpreting $\boldsymbol{P}$ as the deformation gradient associated with a transformation of reference configuration,

$$\begin{aligned} \boldsymbol{\sigma} &= \boldsymbol{g}(\boldsymbol{F}) = \boldsymbol{g}(\boldsymbol{F}\boldsymbol{P}) = \boldsymbol{g}^*(\boldsymbol{F}) \\ &= \boldsymbol{g}^*(\boldsymbol{F}^*) = \boldsymbol{g}(\boldsymbol{F}^*\boldsymbol{P}) = \boldsymbol{g}(\boldsymbol{F}^*), \end{aligned} \tag{27}$$

successive use being made of equations (4), (26), (25), (24), (25) and (26). The first, third, fourth and sixth of the equalities (27) together show that in calculating the stress at the representative particle $\mathscr{X}$ it matters not which of the response functions $\boldsymbol{g}$ and $\boldsymbol{g}^*$ nor which of the deformation gradients $\boldsymbol{F}$ and $\boldsymbol{F}^*$ is used: in short, the stress response of the material does not discriminate between the reference configurations B_r and B^*_r.

Evidently the set G of invertible tensors $\boldsymbol{H}$ possessing the property

$$g(\boldsymbol{AH}) = g(\boldsymbol{A}) \quad \forall \text{ invertible } \boldsymbol{A} \tag{28}$$

provides a characterization of the symmetry of the elastic material under consideration in relation to the reference configuration B_r. If $\boldsymbol{H}_1$ and $\boldsymbol{H}_2$ are elements of G then so is $\boldsymbol{H}_1\boldsymbol{H}_2$ since

$$g(\boldsymbol{AH}_1\boldsymbol{H}_2) = g(\boldsymbol{AH}_1) = g(\boldsymbol{A}).$$

Also, if $\boldsymbol{H} \in$ G and $\boldsymbol{A}$ is invertible, $\boldsymbol{AH}^{-1}$ is an invertible tensor and so $g(\boldsymbol{A}) = g(\boldsymbol{AH}^{-1})$, implying that $\boldsymbol{H}^{-1} \in$ G. These properties ensure that G has the structure of a group:[3] it is referred to as the *symmetry group* of the material relative to B_r.

Problem 6 In the situation discussed above, show that the symmetry group of the material relative to B_r^* is given by

$$\mathrm{G}^* = \boldsymbol{P}\mathrm{G}\boldsymbol{P}^{-1}, \tag{29}$$

$\boldsymbol{P}$ being the deformation gradient associated with the change of reference configuration $B_r \to B_r^*$.

Solution. Let $\boldsymbol{H}$ belong to G. Then (28) holds and it is permissible to replace $\boldsymbol{A}$ by $\boldsymbol{AP}$ in this equation. On using the definition (25) in the form $g(\boldsymbol{A}) = g^*(\boldsymbol{AP}^{-1})$ we then find that

$$g^*(\boldsymbol{APHP}^{-1}) = g^*(\boldsymbol{A}) \quad \forall \text{ invertible } \boldsymbol{A}.$$

Thus if $\boldsymbol{H} \in$ G, $\boldsymbol{PHP}^{-1}$ is an element of G* and the relation (29) is established. [This result is known as *Noll's rule.*]

It is an immediate consequence of equation (29) that if $\boldsymbol{P}$ is a spherical tensor, G* = G. This means that an elastic material has identical symmetry groups relative to two reference configurations which are linked by a dilatation (i.e. a triaxial strain in which the principal stretches are all equal), combined possibly with the central inversion represented by $-\boldsymbol{I}$. Since an arbitrary invertible tensor can be expressed as the product of a unimodular tensor[4] (i.e. a tensor whose determinant is ± 1) and a spherical tensor, it suffices from now on to consider changes of reference configura-

[3] See, for example, L. Mirsky, *An Introduction to Linear Algebra* (Oxford, Clarendon Press, 1955), p. 263.

[4] $\boldsymbol{A} = \{\pm|\det \boldsymbol{A}|^{-1/3}\boldsymbol{A}\}\{\pm|\det \boldsymbol{A}|^{1/3}\boldsymbol{I}\}$.

tion for which the associated deformation gradient is a member of the unimodular group U. It follows from equations (3.6) and (23) that the density is unaltered by such a transformation.[5] The set of all unimodular tensors $\boldsymbol{H}$ satisfying (28) is called the *isotropy group* of the material relative to B_r. If $\boldsymbol{H} \in \mathsf{U}$ and $\boldsymbol{P}$ is an arbitrary invertible tensor, then $\boldsymbol{PHP}^{-1} \in \mathsf{U}$. Thus, as a further consequence of Noll's rule, an isotropy group maps into an isotropy group under a change of reference configuration. In particular, since $\mathsf{U} = \boldsymbol{P}\mathsf{U}\boldsymbol{P}^{-1}$, an isotropy group which coincides with the full unimodular group in relation to some B_r has this property relative to all reference configurations.

Problem 7 Prove that an elastic material whose isotropy group coincides with the unimodular group is a compressible inviscid fluid.

Solution. Relative to an arbitrary reference configuration B_r equation (28) holds for all $\boldsymbol{H} \in \mathsf{U}$. Hence, making the particular choice $\boldsymbol{H} = |J|^{1/3}\boldsymbol{F}^{-1}$,

$$\boldsymbol{g}(\boldsymbol{F}) = \boldsymbol{g}(|J|^{1/3}\boldsymbol{I}). \tag{A}$$

The substitution of (A) into the restriction (17) imposed on $\boldsymbol{g}$ by the principle of objectivity leads to the condition

$$\boldsymbol{Q}\boldsymbol{g}(|J|^{1/3}\boldsymbol{I}) = \boldsymbol{g}(|J|^{1/3}\boldsymbol{I})\boldsymbol{Q} \quad \forall \text{ proper orthogonal } \boldsymbol{Q},$$

and it follows from Exercise 1.11 (p. 49) that $\boldsymbol{g}(|J|^{1/3}\boldsymbol{I})$ is a scalar function of $|J|$ times the identity tensor $\boldsymbol{I}$. In view of equation (3.6) a function of $|J|$ can be expressed in terms of the density ρ. Thus equation (4) reduces to the constitutive relation (2) characterizing an inviscid fluid.

[When the isotropy group coincides with U the symmetry group relative to an arbitrary reference configuration is the set of all invertible tensors. A compressible inviscid fluid is therefore an elastic material with maximal symmetry possessing no preferred configuration. These properties are plainly consistent with intuitive notions of fluidity.]

[5] It is not assumed in this discussion that B_r and B_r^* are similar configurations. Note, therefore, that when the two configurations involved are dissimilar, equation (2.22) becomes $dv = -J\,dV$ and the referential equation of continuity is, in place of (3.6), $\rho = -J^{-1}\rho_r$.

An elastic material is said to be *isotropic* if it possesses a reference configuration in relation to which the isotropy group contains the full orthogonal group O (i.e. the set of all orthogonal tensors). Such a configuration is called an *undistorted state* of the material. The stress response of an isotropic elastic material is therefore unaffected by a change of reference configuration consisting of an arbitrary rotation of an undistorted state combined perhaps with the central inversion $-\boldsymbol{I}$, and it is in this sense that the material may be said to exhibit no preferred directions. In the case of the elastic fluid considered in Problem 7 every configuration is an undistorted state.

Suppose now that the constitutive equation (4) refers to an isotropic elastic material and that the reference configuration B_r involved in the definitions of $\boldsymbol{g}$ and $\boldsymbol{F}$ is an undistorted state of the material. Then it follows from (28) that

$$\boldsymbol{g}(\boldsymbol{F}\boldsymbol{Q}) = \boldsymbol{g}(\boldsymbol{F}) \quad \forall \text{ orthogonal } \boldsymbol{Q}. \tag{30}$$

The result of replacing $\boldsymbol{F}$ by its left polar decomposition $\boldsymbol{V}\boldsymbol{R}$ and setting $\boldsymbol{Q} = \boldsymbol{R}^{\mathrm{T}}$ on the left-hand side of (30) is $\boldsymbol{g}(\boldsymbol{F}) = \boldsymbol{g}(\boldsymbol{V})$, and since $\boldsymbol{V}$ is the unique positive definite square root of the left Cauchy–Green strain tensor $\boldsymbol{B}$, equation (4) reduces to

$$\boldsymbol{\sigma} = \boldsymbol{h}(\boldsymbol{B}), \tag{31}$$

where $\boldsymbol{h}$ is a symmetric tensor-valued function. The stress in an isotropic elastic material is therefore wholly determined by the stretch measured from an undistorted state. To this extent (31) is a specialization of the reduced constitutive equation (22), but whereas (22) results from the application to equation (4) of the principle of objectivity, (31) stems directly from the material symmetry condition (28). The restriction placed on the response function $\boldsymbol{h}$ by the principle of objectivity, found by entering into the condition (17) the identity $\boldsymbol{g}(\boldsymbol{F}) = \boldsymbol{h}(\boldsymbol{F}\boldsymbol{F}^{\mathrm{T}})$, is

$$\boldsymbol{h}(\boldsymbol{Q}\boldsymbol{B}\boldsymbol{Q}^{\mathrm{T}}) = \boldsymbol{Q}\boldsymbol{h}(\boldsymbol{B})\boldsymbol{Q}^{\mathrm{T}} \quad \forall \text{ proper orthogonal } \boldsymbol{Q}. \tag{32}$$

A function $\boldsymbol{k}$ from the set of all symmetric tensors into itself is said to be *isotropic* if, for all symmetric tensors $\boldsymbol{S}$ and for all orthogonal $\boldsymbol{Q}$, $\boldsymbol{k}(\boldsymbol{Q}\boldsymbol{S}\boldsymbol{Q}^{\mathrm{T}}) = \boldsymbol{Q}\boldsymbol{k}(\boldsymbol{S})\boldsymbol{Q}^{\mathrm{T}}$. It can be shown[6] that such a function admits the representation

[6] For a proof of this representation theorem see C. Truesdell and W. Noll, *The Non-Linear Field Theories of Mechanics* (Berlin etc., Springer, 1965), p. 32.

$$k(S) = \eta_0 I + \eta_1 S + \eta_2 S^2, \tag{33}$$

where η_0, η_1 and η_2 are scalar functions of the principal invariants of S. In view of (32) the response function h is isotropic[7] and the constitutive equation (31) specifying the stress response of an isotropic elastic material relative to an undistorted state can accordingly be put into the form

$$\sigma = \phi_0 I + \phi_1 B + \phi_2 B^2, \tag{34}$$

where ϕ_0, ϕ_1 and ϕ_2 are scalar functions of the principal invariants of B. It follows at once from equation (34) that the stress and strain tensors σ and B are coaxial in every configuration of an isotropic elastic material and that the stress in an undistorted state of such a material is spherical.

Problem 8 If an isotropic elastic body possesses two undistorted states, B_r and B_r^*, which are similar to one another, prove that the deformation $B_r \to B_r^*$ is a combination of a rotation and a dilatation.

Solution. The isotropy groups of the material relative to B_r and B_r^* each coincide with O and it is therefore a consequence of Noll's rule (29) that PQP^{-1} is orthogonal for all $Q \in$ O, P being the deformation gradient associated with the mapping $B_r \to B_r^*$. Hence, recalling equation $(1.76)_1$,

$$(P^{-1})^T Q^T P^T = PQ^T P^{-1} \quad \forall \text{ orthogonal } Q,$$

from which we see, on premultiplying each side by P^T and postmultiplying by P, that $P^T P$ commutes with every orthogonal tensor and is consequently a positive definite spherical tensor (see Problem 1.11, p. 27 and Exercise 1.11, p. 49). Now, according to the polar decomposition theorem. P is expressible in the form RU, and because B_r and B_r^* are similar configurations the orthogonal tensor R is proper. Furthermore, $U = (P^T P)^{1/2}$ which has been shown to be spherical, so the proof is complete.

An elastic material is called a *solid* if it has a configuration B_r relative to which the isotropy group consists entirely of orthogonal tensors. Physical motivation for this definition comes from the

[7] Since Q occurs quadratically in (32) the condition obviously holds for all orthogonal Q.

experience that the stress response of a solid material is generally altered by a change of shape (as represented by a non-orthogonal transformation of reference configuration). We again describe the preferred configuration B_r as an undistorted state of the material, this term retaining its previous meaning when the solid happens to be isotropic, that is when it has maximal symmetry. An undistorted state of an elastic material in which the stress is zero is called a *natural state*. An elastic solid is said to be *anisotropic* if its isotropy group relative to an undistorted state is a proper subgroup of O.

The theory of material symmetry, as developed in this section for elastic materials, is centred on the idea of invariance under changes of reference configuration, but, as pointed out in Section 2, constitutive equations for viscous and inviscid fluids can be formulated without introducing a reference configuration. It is therefore meaningful to attribute to all such fluids the maximal symmetry which has been shown in Problem 7 to be a characteristic of inviscid fluids. In particular, a viscous fluid is isotropic in every configuration, a property now seen to be guaranteed by the principle of objectivity through the restriction (20) on the response function $\boldsymbol{f}$, and a second application of the representation theorem (33) allows the constitutive equation (19) to be further reduced to

$$\boldsymbol{\sigma} = (-p + \nu_0)\boldsymbol{I} + \nu_1\boldsymbol{D} + \nu_2\boldsymbol{D}^2. \tag{35}$$

Here p is a function of ρ, and ν_0, ν_1 and ν_2 are functions of ρ and the principal invariants of $\boldsymbol{D}$, ν_0 vanishing when $\boldsymbol{D} = \mathbf{O}$.

Equations (34) and (35) are the most general forms of the constitutive equations of an isotropic elastic material and a viscous fluid which are compatible with the principle of objectivity. Equation (35) was first obtained by Reiner and Rivlin, and viscous fluids of the type considered in this chapter are known as *Reiner–Rivlin fluids*.

A *Newtonian viscous fluid* is a Reiner–Rivlin fluid for which the dependence of the extra stress on the stretching tensor is linear. In this case equation (35) simplifies to

$$\boldsymbol{\sigma} = \{-p + (\kappa - \tfrac{2}{3}\mu)\,\mathrm{tr}\,\boldsymbol{D}\}\boldsymbol{I} + 2\mu\boldsymbol{D}, \tag{36}$$

where p, κ and μ are functions of ρ. κ is called the *bulk viscosity* and μ the *shear viscosity* (or simply the *viscosity*) of the fluid. The most common naturally-occurring fluids, air and water, as well as other gases and liquids (such as mercury and some light oils) are well represented by the Newtonian constitutive equation. But there is abundant evidence of fluids behaving in a manner which cannot be explained by the theory of Newtonian viscous flow, examples arising in motions of emulsions (such as paints), polymer solutions and melts, pastes and glues, and blood. At the time of its discovery (1945–48) it was hoped that the Reiner–Rivlin constitutive equation would provide a theoretical basis for studying the dynamics of such fluids, but experimental evidence to date suggests that the only known Reiner–Rivlin fluids are those which are Newtonian. Much of the recent work on non-Newtonian flow is therefore concerned with characterizations of viscosity more general than equation (3).

6 INTERNAL CONSTRAINTS

Up to now it has been tacitly supposed that a material body can be made to execute any (smooth) motion by the application of suitable forces. But considerable importance attaches to the theoretical study of fluids and solid materials which suffer some restriction on the class of deformations which they are able to undergo. Such a limitation is called an *internal* (or *kinematic*) *constraint* and the most important examples, to be considered here, are represented by equations of the form

$$\zeta(\boldsymbol{F}) = 0, \tag{37}$$

where ζ is a scalar-valued function. The requirement that internal constraints shall be objective leads, via equation (12), to the requirement

$$\zeta(\boldsymbol{Q}\boldsymbol{F}) = \zeta(\boldsymbol{F}) \quad \forall \text{ proper orthogonal } \boldsymbol{Q}. \tag{38}$$

On replacing $\boldsymbol{F}$ by its right polar decomposition $\boldsymbol{R}\boldsymbol{U}$ and setting $\boldsymbol{Q} = \boldsymbol{R}^{\mathrm{T}}$ on the left-hand side of (38) we find that ζ depends upon the deformation gradient through $\boldsymbol{U}$ and hence through the right Cauchy–Green strain tensor $\boldsymbol{C}$. Equation (37) is therefore superseded by the objective condition

$$\lambda(\boldsymbol{C}) = 0. \tag{39}$$

In order to accommodate its motion to an internal constraint a material body $\mathscr{B}$ must be able to bring appropriate contact forces into play, and the constitutive equation governing its stress response must be such as to allow these forces to act. The principle of determinism, as stated in Section 1, must therefore be modified and this is done by assuming that the part of the stress arising through the operation of the constraint does no work and is not determined by the motion. In the case of a constrained body the basic constitutive statement corresponding to (1) therefore reads as follows.

*The stress at the point x at time t is uniquely determined by the values at (x, t) of fields derived from the motion of $\mathscr{B}$ to within a symmetric tensor **N** which is not determined in this manner by the motion of $\mathscr{B}$ and which does no work in any motion compatible with the constraints.*

Recalling the interpretation of stress power given in Section 3.5, the condition that (39) shall be a workless constraint takes the form

$$\operatorname{tr}(\boldsymbol{N}\boldsymbol{D}) = 0 \tag{40}$$

for all allowable motions. In order to arrive at a relationship between $\boldsymbol{N}$ and λ we utilize equation (39) in the differentiated form $\dot{\lambda} = 0$. With respect a rectangular Cartesian system $(\mathrm{O}, \boldsymbol{E})$ of referential coordinates,

$$\dot{\lambda} = (\partial\lambda/\partial C_{\pi\rho})\dot{C}_{\pi\rho} = \operatorname{tr}(\lambda_C \dot{\boldsymbol{C}}), \tag{41}$$

where λ_C is the symmetric tensor given, in component form, by[8]

$$\lambda_C = (\partial\lambda/\partial C_{\pi\rho})\boldsymbol{E}_\pi \otimes \boldsymbol{E}_\rho. \tag{42}$$

Now $\dot{\boldsymbol{C}} = 2\boldsymbol{F}^{\mathrm{T}}\boldsymbol{D}\boldsymbol{F}$ (see Exercise 1 (p. 160)), whence, using the general property $(1{\cdot}53)_2$ of the trace operator,

$$\operatorname{tr}(\boldsymbol{F}\lambda_C\boldsymbol{F}^{\mathrm{T}}\boldsymbol{D}) = 0. \tag{43}$$

The condition (40) holds for all stretching tensors $\boldsymbol{D}$ which satisfy (43) and it follows from Problem 1·12 (p. 27) that $\boldsymbol{N}$ is a scalar multiple of $\boldsymbol{F}\lambda_C\boldsymbol{F}^{\mathrm{T}}$:

$$\boldsymbol{N} = \alpha\boldsymbol{F}\lambda_C\boldsymbol{F}^{\mathrm{T}}. \tag{44}$$

The *constraint stress* $\boldsymbol{N}$ is therefore determined, to within a scalar multiplier, by the form (39) of the constraint. If $\mathscr{B}$ is subject to a number of internal constraints,[9] given by

[8] It is understood here that the dependence of λ on the strain components $C_{\alpha\beta}$ is expressed symmetrically.

[9] Since equation (39) effectively determines one component of the strain tensor $\boldsymbol{C}$, the maximum number of independent constraints of this type is six.

$$\lambda_\Gamma(C) = 0, \quad \Gamma = 1, 2, \ldots, n,$$

the foregoing argument again applies with λ replaced by an arbitrary linear combination of $\lambda_1, \lambda_2, \ldots, \lambda_n$. The constraint stress is therefore given by

$$N = \sum_{\Gamma=1}^{n} \alpha_\Gamma \boldsymbol{F}(\lambda_\Gamma)_C \boldsymbol{F}^{\mathrm{T}}. \tag{45}$$

We have already met one example of an internal constraint in Chapter 3: an incompressible material is one which can perform only isochoric motions and since, in this case, $J = (\det \boldsymbol{C})^{1/2} = 1$, the appropriate form of equation (39) is

$$\lambda(\boldsymbol{C}) = \det \boldsymbol{C} - 1 = 0. \tag{46}$$

The rule (1.40) for differentiating a determinant gives

$$\dot{\lambda} = (\det \boldsymbol{C}) \operatorname{tr} (\dot{\boldsymbol{C}}\boldsymbol{C}^{-1}) = \operatorname{tr} (\boldsymbol{C}^{-1}\dot{\boldsymbol{C}})$$

and so, by comparison with (41), $\lambda_C = \boldsymbol{C}^{-1} = \boldsymbol{F}^{-1}(\boldsymbol{F}^{-1})^{\mathrm{T}}$. The constraint stress is therefore given by equation (44) as

$$\boldsymbol{N} = -p\boldsymbol{I},$$

the disposable scalar, here denoted by $-p$, having an obvious interpretation as a pressure.

For an incompressible isotropic elastic material the constitutive equation (34) is replaced by

$$\boldsymbol{\sigma} = -p\boldsymbol{I} + \phi_1 \boldsymbol{B} + \phi_2 \boldsymbol{B}^2, \tag{47}$$

where ϕ_1 and ϕ_2 are functions of I_B and II_B only (since $III_B = J^2 = 1$) and the spherical term $\phi_0 \boldsymbol{I}$ has been absorbed into the constraint stress.[10] Equation (47) provides accurate models of the behaviour of natural and synthetic rubbers.

The very low compressibility displayed by elastomeric materials at conventional stress levels is also a fundamental property of liquids, and in consequence the constitutive equation representing an incompressible Newtonian fluid is the basis of extensive developments in the theory of viscous flow. Appropriately modifying equation (36) and remembering, from equation (2.32), that $\operatorname{tr} \boldsymbol{D} = 0$ in an

[10] That no loss of generality is involved in this step will be better appreciated after a study of Problem 15 (p. 158) in which the calculation of the pressure p is exemplified

isochoric motion we obtain this equation in the simple form

$$\boldsymbol{\sigma} = -p\boldsymbol{I} + 2\mu\boldsymbol{D}, \tag{48}$$

where the viscosity μ is now a constant. On substituting (48) into the equation of motion (3.23), and again using the spatial continuity equation in the form (2.32), we arrive at the *Navier–Stokes equation*

$$\rho\boldsymbol{a} = -\operatorname{grad} p + \mu \,\text{☼}\boldsymbol{v} + \rho\boldsymbol{b} \tag{49}$$

in which ☼$(\cdot) = \operatorname{div}\{\operatorname{grad}(\cdot)\}^{\mathrm{T}}$ is the vector Laplacian operator in the spatial description.

Equations (2.32) and (49) supply four scalar relations between p and the components of $\boldsymbol{v}$, the pressure entering the list of dependent variables in place of the density which is prescribed by the constraint. Some solutions of these equations are given in Section 7 below. In the case of an incompressible inviscid fluid the whole of the stress is devoted to maintaining the constraint and the equations governing the pressure and velocity are (2.32) and (3.25). A particular solution has been discussed in Problem 3.7 (p. 101).

Problem 9 Returning to Problem 2.11 (p. 72), determine the constraint stress associated with the two families of embedded fibres.

Solution. From equations (*C*) in the solution of Problem 2.11 the internal constraints are given by

$$\check{\lambda}_\Gamma(\boldsymbol{C}) = \boldsymbol{L}_\Gamma\cdot(\boldsymbol{C}\boldsymbol{L}_\Gamma) - 1 = 0, \quad \Gamma = 1, 2,$$

where $\boldsymbol{L}_1$ and $\boldsymbol{L}_2$ are unit vectors defining the fibre directions in the undeformed configuration B_r. Introducing a rectangular Cartesian system $(O, \boldsymbol{E})$ of referential coordinates and observing footnote 8 on p. 146 we find that

$$\frac{\partial\lambda_\Gamma}{\partial C_{\alpha\beta}} = \frac{\partial}{\partial C_{\alpha\beta}}\{\tfrac{1}{2}(L_\Gamma)_\pi(L_\Gamma)_\rho(C_{\pi\rho} + C_{\rho\pi})\}$$

$$= \tfrac{1}{2}(L_\Gamma)_\pi(L_\Gamma)_\rho(\delta_{\alpha\pi}\delta_{\beta\rho} + \delta_{\alpha\rho}\delta_{\beta\pi}) = (L_\Gamma)_\alpha(L_\Gamma)_\beta,$$

leading, via the definition (42), to $(\lambda_\Gamma)_C = \boldsymbol{L}_\Gamma \otimes \boldsymbol{L}_\Gamma$. Hence

$$\boldsymbol{F}(\lambda_\Gamma)_C\boldsymbol{F}^{\mathrm{T}} = (\boldsymbol{F}\boldsymbol{L}_\Gamma)\otimes(\boldsymbol{F}\boldsymbol{L}_\Gamma) = \boldsymbol{l}_\Gamma\otimes\boldsymbol{l}_\Gamma,$$

the unit vectors $\boldsymbol{l}_1$ and $\boldsymbol{l}_2$ specifying the fibre orientations in the deformed configuration B, and there follows from equation (45) the expression

$$N = T_1 \boldsymbol{l}_1 \otimes \boldsymbol{l}_1 + T_2 \boldsymbol{l}_2 \otimes \boldsymbol{l}_2$$

for the constraint stress. From Exercise 3.2 (p. 121), we recognize N as a superposition of uniaxial stresses in the directions taken up by the fibres in B. The disposable scalars T_1 and T_2 associated with the constraints therefore have a clear interpretation as the tensions in the fibres belonging to the two families.

7 INCOMPRESSIBLE NEWTONIAN VISCOUS FLUIDS

In this section we consider some simple flows of an incompressible Newtonian viscous fluid for which the governing equations (2.32) and (49) can be solved exactly. The availability of such solutions materially assists the understanding of fundamental concepts in viscous flow theory, in particular the boundary layer and the diffusion of vorticity. The detailed discussion of these topics properly belongs to specialist texts,[11] however, and is not attempted here.

In each of the problems which follow, body forces are assumed absent and the basic equations are referred to a spatial system of rectangular Cartesian coordinates x_1, x_2, x_3. With the aid of equations (1.92) to (1.94) and $(2.12)_3$ we can write these component forms as

$$\frac{\partial v_p}{\partial x_p} = 0, \quad \rho\left(\frac{\partial v_i}{\partial t} + v_p \frac{\partial v_i}{\partial x_p}\right) = -\frac{\partial p}{\partial x_i} + \mu \frac{\partial^2 v_i}{\partial x_p \partial x_p}, \tag{50}$$

ρ being the density and μ the viscosity of the fluid.

Problem 10 Discuss the steady flow of an incompressible Newtonian viscous fluid through a straight pipe, with particular reference to the case in which the cross-section is circular.

[11] See, for example, L. Rosenhead (ed.), *Laminar Boundary Layers* (Oxford, Clarendon Press, 1963), Chapters II and III, and G. K. Batchelor, *An Introduction to Fluid Dynamics* (Cambridge, Cambridge University Press, 1967).

Solution. Taking the 3-direction to be aligned with the generators of the pipe, we seek a solution of equations (50) in which the velocity is everywhere directed along the pipe:[12]

$$v_1 = 0, \quad v_2 = 0, \quad v_3 = v(x_1, x_2). \tag{A}$$

These expressions clearly satisfy the continuity equation $(50)_1$ and on substituting them into the Navier–Stokes equations $(50)_2$ we find that

$$\frac{\partial p}{\partial x_1} = \frac{\partial p}{\partial x_2} = 0, \quad \frac{\partial p}{\partial x_3} = \mu \Delta_1 v, \tag{B}$$

where

$$\Delta_1(\cdot) = \left(\frac{\partial^2}{\partial x_1^2} + \frac{\partial^2}{\partial x_2^2}\right)(\cdot)$$

is the plane scalar Laplacian operator. It follows from (B) that the pressure p is a function of x_3 only and that $\mathrm{d}p/\mathrm{d}x_3$ does not depend upon x_3. Hence

$$p = p_0 - Gx_3, \tag{C}$$

where p_0 and G are constants, and the steady linear flow described by equations (A) is seen to require a constant pressure gradient for its maintenance. If $G > 0$, p decreases as x_3 increases and the fluid is expected to move in the positive 3-direction.

Equations $(B)_3$ and (C) now yield the Poisson equation

$$\Delta_1 v = -G/\mu \tag{D}$$

which is to be solved in the region of the (x_1, x_2)-plane occupied by a cross-section of the pipe subject to a boundary condition holding on the perimeter c of this region.[13] The appropriate requirement is the *no-slip condition*, stating that, at an interface between a viscous fluid and a solid body, no relative motion of the two materials can occur. Thus

$$v = 0 \text{ on } c. \tag{E}$$

The boundary-value problem consisting of equations (D) and

[12] Equation $(A)_3$ could be replaced by $v_3 = v(x_1, x_2, x_3)$, but the dependence of v on x_3 would then be excluded by the continuity equation.

[13] c consists of one or more simple closed curves.

(E) has a unique bounded solution. For the case in which c is the circle given by $x_1^2 + x_2^2 = a^2$, the solution is easily shown to be

$$v = \tfrac{1}{4}(G/\mu)(a^2 - r^2),$$

where $r = (x_1^2 + x_2^2)^{1/2}$ is distance from the axis of the pipe. The velocity distribution over the cross-section of the pipe in this case is parabolic, the maximum speed being $Ga^2/4\mu$ at the axis. The volume flux of fluid through the pipe is given by *Poiseuille's formula*

$$F = \int_0^a 2\pi r v_3 \, dr = \tfrac{1}{8}\pi G a^4/\mu.$$

Problem 11 An incompressible Newtonian viscous fluid fills the semi-infinite region on one side of a flat rigid plate. Ignoring transients, investigate the vibratory motion of the fluid set up in response to a rectilinear oscillation of the plate in its own plane.

Solution. Placing the origin of coordinates in the plane of the plate with the 2-direction directed normally into the fluid and the 1-direction aligned with the motion of the plate, we assume that the particle velocity is everywhere in the 1-direction and depends only upon distance x_2 from the plate and the time t. Thus

$$v_1 = v(x_2, t), \quad v_2 = 0, \quad v_3 = 0. \tag{A}$$

Equation $(50)_1$ is satisfied regardless of the form of the function v and the Navier–Stokes equations $(50)_2$ hold if

$$\frac{\partial p}{\partial x_1} = \mu \frac{\partial^2 v}{\partial x_2^2} - \rho \frac{\partial v}{\partial t}, \quad \frac{\partial p}{\partial x_2} = \frac{\partial p}{\partial x_3} = 0. \tag{B}$$

These relations imply that p depends only upon x_1 and t and that $\partial p/\partial x_1$ is independent of x_1. The most general form of the pressure distribution consistent with the assumed velocity field is therefore

$$p = p_0(t) - G(t)x_1.$$

We are concerned, however, with the situation in which the fluid flow is driven entirely by the movement of the plate, so the pressure gradient G is taken to be zero. It then follows from $(B)_1$ that v satisfies the one-dimensional diffusion equation

$$\nu \frac{\partial^2 v}{\partial x_2^2} = \frac{\partial v}{\partial t}, \tag{C}$$

wherein $\nu = \mu/\rho$ is the *kinematic viscosity* of the fluid. By virtue of the no-slip condition, v coincides at $x_2 = 0$ with the prescribed velocity of the plate, and in order that far-distant fluid shall be insensible to boundary motions we require that

$$v \to 0 \text{ as } x_2 \to \infty \text{ for all } t. \tag{D}$$

If, in addition, the fluid is assumed to be stationary at $t = 0$ we have an initial-boundary-value problem for v possessing a unique solution.

In the special case in which the plate vibrates with angular frequency ω and velocity amplitude V the boundary condition can be expressed in the complex form

$$v = V\mathrm{e}^{\mathrm{i}\omega t} \text{ at } x_2 = 0. \tag{E}$$

And when the transient disturbance associated with the commencement of the flow is ignored, we discard the initial condition and seek a solution of equation (C) in the form

$$f(x_2)\,\mathrm{e}^{\mathrm{i}\omega t}. \tag{F}$$

The resulting ordinary differential equation for f is

$$f''(x_2) = (\mathrm{i}\omega/\nu)f(x_2) = \{(1 + \mathrm{i})k\}^2 f(x_2), \tag{G}$$

where $k = (\omega/2\nu)^{1/2}$ and we note that this constant has the dimension $(\text{length})^{-1}$. The disposable constants in the general solution of (G) are fixed by the conditions (D) and (E). The solution of equation (C) provided by the real part of (F) is then found to be

$$v = V\mathrm{e}^{-kx_2} \cos(\omega t - kx_2), \tag{H}$$

the corresponding plate velocity being $V \cos \omega t$.

Equations (A) and (H) show that the amplitude of the oscillation performed by the particles in planes parallel to the plate decays exponentially as x_2 increases. Thus the fluid motion is essentially confined to a layer adjoining the plate, the more distant fluid behaving as if the plate were at rest. The thickness of the boundary layer is of order $k^{-1} = (2\nu/\omega)^{1/2}$, indicating that a slowly vibrating plate in a highly viscous medium drags a large amount of fluid along with it, while a plate oscillating rapidly in a fluid of low viscosity causes no appreciable motion outside a thin boundary layer.

The components of vorticity derived from equations (A) and (H) are

$$\omega_1 = 0, \quad \omega_2 = 0, \quad \omega_3 = 2^{1/2} V k \, e^{-kx_2} \cos(\omega t - kx_2 + \tfrac{1}{4}\pi),$$

and the non-zero component ω_3 satisfies the same diffusion equation as v_1. Vorticity evidently originates at the moving boundary and diffuses into the fluid.

Problem 12 Prove that the converging circulatory motion considered in Problem 2.15 (p. 81) can be supported, without the aid of body forces, by an incompressible Newtonian viscous fluid. Compare the distribution of vorticity associated with this solution with the results obtained in the earlier problem for a circulation preserving motion.

Solution. The components of velocity and vorticity in the given motion are

$$v_1 = -\tfrac{1}{2}\alpha x_1 - f(r)x_2, \quad v_2 = -\tfrac{1}{2}\alpha x_2 + f(r)x_1, \quad v_3 = \alpha x_3, \tag{A}$$

and

$$\omega_1 = 0, \quad \omega_2 = 0, \quad \omega_3 = \omega(r) = rf'(r) + 2f(r), \tag{B}$$

where $r = (x_1^2 + x_2^2)^{1/2}$ and α is a positive constant. As the motion is already known to be isochoric our task is to show that the function f and the pressure distribution can be so chosen that the velocity components (A) satisfy the Navier–Stokes equations $(50)_2$.

The three relations obtained by entering (A) into equations $(50)_2$ can be written as

$$\left.\begin{aligned} \frac{\partial p}{\partial x_1} = -\chi(r)x_1 - \psi(r)x_2, \quad \frac{\partial p}{\partial x_2} = -\chi(r)x_2 + \psi(r)x_1, \\ \frac{\partial p}{\partial x_3} = -\rho\alpha^2 x_3, \end{aligned}\right\} \tag{C}$$

where

$$\chi(r) = \tfrac{1}{4}\rho\alpha^2 - \rho\{f(r)\}^2, \quad \psi(r) = \mu r^{-1}\omega'(r) + \tfrac{1}{2}\rho\alpha\omega(r), \tag{D}$$

and ω is given by $(B)_3$. The integrability conditions

$$\partial^2 p/\partial x_i \partial x_j = \partial^2 p/\partial x_j \partial x_i,$$

which suffice for the existence of a function p satisfying (C), reduce to the single equation

$$r\psi'(r) + 2\psi(r) = 0,$$

the only bounded solution of which is $\psi = 0$. Equation $(D)_2$ now yields a first-order differential equation for the vorticity ω and, with the use of $(D)_1$, we can effect the integration of equations (C) in the form

$$p = p_0 - \tfrac{1}{8}\rho\alpha^2(r^2 + 4x_3^2) + \rho \int^r \{f(s)\}^2 s \, ds,$$

p_0 being a constant. The solution of the equation for ω is

$$\omega(r) = \omega_0 \exp(-\tfrac{1}{4}\alpha r^2/\nu), \tag{E}$$

where $\nu = \mu/\rho$ is the kinematic viscosity and ω_0 is an additional constant, and it remains only to observe that f can be determined by integration from equations $(B)_3$ and (E).

Equation (E) shows that the vorticity is everywhere finite and strongly concentrated towards a central core with radius of order $2(\nu/\alpha)^{1/2}$. Outward diffusion of vorticity from this inner region is apparent, but, to a close approximation, the motion of the surrounding fluid is irrotational and uninfluenced by viscosity. In contrast, the vorticity in the circulation preserving motion considered in Problem 2.15 (p. 81) is singular on the 3-axis. Since this is a possible motion under zero body forces of an incompressible *inviscid* fluid, the smoothing out of the line vortex and the diffusion of vorticity are directly attributable to the presence of viscosity.

Whereas in inviscid flow the effects associated with the convection and stretching of vortex lines balance one another, in viscous flow these processes are together in balance with the diffusion of vorticity and vortex lines are generally in motion relative to the material.

8 ISOTROPIC ELASTIC MATERIALS

The problems in this concluding section illustrate some elementary but instructive aspects of the non-linear theory of elasticity for isotropic materials. We see from the constitutive equation (4) that when a compressible elastic body is subjected to a homogeneous deformation the stress in the current configuration is uniform, and it then follows from the equilibrium equation (3.26) that the body can rest permanently in such a state without the intervention of body forces, that is under the action of surface tractions only. Problems 13 and 14 are concerned with situations of this kind. In Problem 15 a simple motion of an incompressible elastic body is studied.

Problem 13 Calculate the stress in a compressible isotropic elastic body subjected to the state of simple shear discussed in Problem 2.10 (p. 69).

Solution. This is a homogeneous isochoric deformation in which the left Cauchy–Green strain tensor is given by

$$\boldsymbol{B} = (1 + \gamma^2)\boldsymbol{l} \otimes \boldsymbol{l} + \boldsymbol{m} \otimes \boldsymbol{m} + \boldsymbol{n} \otimes \boldsymbol{n} + \gamma(\boldsymbol{l} \otimes \boldsymbol{m} + \boldsymbol{m} \otimes \boldsymbol{l}), \quad \text{(A)}$$

where γ is the amount of shear and the definition of the base vectors $\boldsymbol{l}$, $\boldsymbol{m}$, $\boldsymbol{n}$ is indicated in Figure 2.5. Thus

$$\boldsymbol{B}^2 = (1 + 3\gamma^2 + \gamma^4)\boldsymbol{l} \otimes \boldsymbol{l} + (1 + \gamma^2)\boldsymbol{m} \otimes \boldsymbol{m} + \boldsymbol{n} \otimes \boldsymbol{n} \\ + \gamma(2 + \gamma^2)(\boldsymbol{l} \otimes \boldsymbol{m} + \boldsymbol{m} \otimes \boldsymbol{l}) \qquad \text{(B)}$$

and the constitutive equation (34) yields the expression

$$\boldsymbol{\sigma} = \{\phi_0 + \phi_1(1 + \gamma^2) + \phi_2(1 + 3\gamma^2 + \gamma^4)\}\boldsymbol{l} \otimes \boldsymbol{l} \\ + \{\phi_0 + \phi_1 + \phi_2(1 + \gamma^2)\}\boldsymbol{m} \otimes \boldsymbol{m} + (\phi_0 + \phi_1 + \phi_2)\boldsymbol{n} \otimes \boldsymbol{n} \\ + \gamma\{\phi_1 + \phi_2(2 + \gamma^2)\}(\boldsymbol{l} \otimes \boldsymbol{m} + \boldsymbol{m} \otimes \boldsymbol{l}) \qquad \text{(C)}$$

for the stress. From (A) and (B) the principal invariants of $\boldsymbol{B}$ are found to be

$$I_{\boldsymbol{B}} = II_{\boldsymbol{B}} = 3 + \gamma^2, \quad III_{\boldsymbol{B}} = 1,$$

Exercise 1.2 (p. 47) providing a convenient way of calculating $II_{\boldsymbol{B}}$. In (C), ϕ_0, ϕ_1 and ϕ_2 may therefore be regarded as functions of γ^2.

The components of stress relative to the chosen basis are obtained from equation (C) in the form

$$\left.\begin{aligned} \sigma_{(l)} &= \boldsymbol{l}.(\boldsymbol{\sigma}\boldsymbol{l}) = \phi_0 + \phi_1(1+\gamma^2) + \phi_2(1+3\gamma^2+\gamma^4), \\ \sigma_{(m)} &= \boldsymbol{m}.(\boldsymbol{\sigma}\boldsymbol{m}) = \phi_0 + \phi_1 + \phi_2(1+\gamma^2), \\ \sigma_{(n)} &= \boldsymbol{n}.(\boldsymbol{\sigma}\boldsymbol{n}) = \phi_0 + \phi_1 + \phi_2, \end{aligned}\right\} \quad \text{(D)}$$

$$\boldsymbol{m}.(\boldsymbol{\sigma}\boldsymbol{n}) = 0, \quad \boldsymbol{n}.(\boldsymbol{\sigma}\boldsymbol{l}) = 0, \quad \boldsymbol{l}.(\boldsymbol{\sigma}\boldsymbol{m}) = \gamma\{\phi_1 + \phi_2(2+\gamma^2)\}. \quad \text{(E)}$$

These formulae display the striking fact that normal stresses as well as a shear stress are needed to produce a state of simple shear in an isotropic elastic material.[14] The central importance of normal stresses in simple shear is further emphasized by the relation

$$\sigma_{(l)} - \sigma_{(m)} = \gamma \boldsymbol{l}.(\boldsymbol{\sigma}\boldsymbol{m}) \quad \text{(F)}$$

which follows from $(\mathrm{D})_{1,2}$ and $(\mathrm{E})_3$. This result is remarkable in being independent of the functions ϕ_0, ϕ_1 and ϕ_2 specifying the stress response of the material, and it is consequently referred to as a *universal relation.* Equation (F) shows in the most direct fashion that the shear stress $\boldsymbol{l}.(\boldsymbol{\sigma}\boldsymbol{m})$ arises from the normal stress difference $\sigma_{(l)} - \sigma_{(m)}$ and, moreover, that it is so produced in exactly the same way in every isotropic elastic material. It also follows from (F) that the normal stresses $\sigma_{(l)}$ and $\sigma_{(m)}$ are unequal, a property known as the *Poynting effect.*

Problem 14 It is conjectured that the constitutive equation describing the stress response of a certain class of foam rubbers is

$$\boldsymbol{\sigma} = III_B^{-3/2}[\{\psi(III_B) - \beta II_B\}\boldsymbol{I} + (\alpha III_B + \beta I_B)\boldsymbol{B} - \beta\boldsymbol{B}^2],$$

where α and β are constants. In an experiment a cylindrical test-piece is given a homogeneous deformation in which the principal stretches are Λ, λ, λ, the current stretch axis associated with Λ being aligned with the generators of the cylinder, and a plot of the transverse stretch λ against the axial stretch Λ on logarithmic scales is found to be a straight line with negative slope. Assuming that the curved boundary of the cylinder is traction-free, deduce the form of the function ψ.

[14] If the normal stresses were all zero it would follow from equations (D) that ϕ_0, ϕ_1 and ϕ_2 all vanish at the positive value of γ^2 in question and hence, from $(\mathrm{E})_3$, that $\boldsymbol{l}.(\boldsymbol{\sigma}\boldsymbol{m}) = 0$. It would therefore be possible to shear the material at zero stress, a degeneracy of no interest.

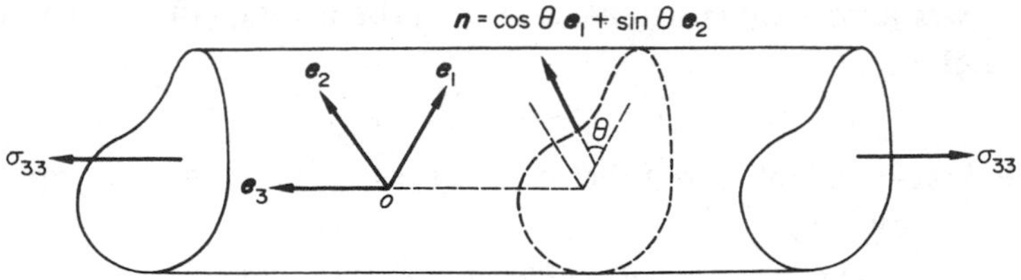

FIGURE 1 Deformed configuration of a cylinder subject to uniaxial extension (Problem 14).

Solution. The orthonormal vectors $\boldsymbol{e}_1$, $\boldsymbol{e}_2$, $\boldsymbol{e}_3$ shown in Figure 1 define the current stretch axes at an arbitrary point in the deformed configuration of the test-piece and the left Cauchy–Green strain tensor is therefore given by

$$\boldsymbol{B} = \lambda^2(\boldsymbol{e}_1 \otimes \boldsymbol{e}_1 + \boldsymbol{e}_2 \otimes \boldsymbol{e}_2) + \Lambda^2 \boldsymbol{e}_3 \otimes \boldsymbol{e}_3 = \lambda^2 \boldsymbol{I} + (\Lambda^2 - \lambda^2)\boldsymbol{e}_3 \otimes \boldsymbol{e}_3.$$

The principal invariants of $\boldsymbol{B}$, obtained with the aid of equations (1.60) to (1.62), are

$$I_{\boldsymbol{B}} = 2\lambda^2 + \Lambda^2, \quad II_{\boldsymbol{B}} = 2\lambda^2\Lambda^2 + \lambda^4, \quad III_{\boldsymbol{B}} = \lambda^4\Lambda^2,$$

and the stress in the cylinder, calculated from the conjectured constitutive equation, is hence

$$\boldsymbol{\sigma} = \lambda^{-6}\Lambda^{-3}[\{\psi(\lambda^4\Lambda^2) + \alpha\lambda^6\Lambda^2 - \beta\lambda^2\Lambda^2\}\boldsymbol{I} + (\alpha\lambda^4\Lambda^2 + \beta\lambda^2)(\Lambda^2 - \lambda^2)\boldsymbol{e}_3 \otimes \boldsymbol{e}_3].$$

The stress vector on the curved boundary of the cylinder is

$$\boldsymbol{t}_{(\boldsymbol{n})} = \boldsymbol{\sigma}^{\mathrm{T}}\boldsymbol{n} = \lambda^{-6}\Lambda^{-3}\{\psi(\lambda^4\Lambda^2) + \alpha\lambda^6\Lambda^2 - \beta\lambda^2\Lambda^2\} \times (\cos\theta\, \boldsymbol{e}_1 + \sin\theta\, \boldsymbol{e}_2),$$

where the unit normal vector $\boldsymbol{n}$ and the angle θ are as shown in Figure 1. Since $\boldsymbol{t}_{(\boldsymbol{n})} = \boldsymbol{0}$ for all values of θ in the interval $[0, 2\pi)$, it follows that

$$\psi(\lambda^4\Lambda^2) + \alpha\lambda^6\Lambda^2 - \beta\lambda^2\Lambda^2 = 0. \tag{A}$$

The information derived from experiments takes the form

$$\ln\lambda = -\nu\ln\Lambda, \quad \text{i.e. } \lambda = \Lambda^{-\nu}, \tag{B}$$

where v is a positive constant, and on substituting (B) into (A) we find that

$$\psi(\Lambda^{2-4v}) = -\alpha\Lambda^{2-6v} + \beta\Lambda^{2-2v}.$$

Since this is an identity in Λ the required result is obtained on replacing Λ^{2-4v} by III_B:

$$\psi(III_B) = -\alpha III_B^{(1-3v)/(1-2v)} + \beta III_B^{(1-v)/(1-2v)}.$$

The axial tension required to maintain the deformation is

$$\sigma_{33} = (\alpha\Lambda + \beta\Lambda^{-1+2v})(\Lambda^{2v} - \Lambda^{-2}).$$

Problem 15 A rubber cylinder which has radius A and length L in its natural state is rotated about its axis of symmetry with constant angular speed ω, the motion being given by

$$x_1 = \lambda^{-1/2}(X_1 \cos \omega t - X_2 \sin \omega t),$$

$$x_2 = \lambda^{-1/2}(X_1 \sin \omega t + X_2 \cos \omega t), \quad x_3 = \lambda X_3,$$

where the referential and spatial coordinates relate to a common rectangular Cartesian system and λ is a positive constant. Verify that the motion is isochoric and find the principal stretches. The rubber is incompressible and may be regarded as a *Mooney material* characterized by the constitutive equation

$$\boldsymbol{\sigma} = -p\boldsymbol{I} + (\alpha + \beta I_B)\boldsymbol{B} - \beta\boldsymbol{B}^2$$

in which α and β are positive constants. Assuming that the curved boundary of the cylinder is traction-free and that no body forces act, determine the pressure p. Assuming further that the resultant forces on the end-faces of the cylinder are zero, obtain an equation from which the length of the spinning cylinder may be calculated.

Solution. The deformation gradient associated with the given motion is

$$\boldsymbol{F} = \lambda^{-1/2}(\boldsymbol{e}_1 \otimes \boldsymbol{e}_1 + \boldsymbol{e}_2 \otimes \boldsymbol{e}_2)\cos \omega t + \lambda \boldsymbol{e}_3 \otimes \boldsymbol{e}_3 - \lambda^{-1/2}(\boldsymbol{e}_1 \otimes \boldsymbol{e}_2 - \boldsymbol{e}_2 \otimes \boldsymbol{e}_1)\sin \omega t,$$

where $\boldsymbol{e}_1, \boldsymbol{e}_2, \boldsymbol{e}_3$, the base vectors of the common coordinate system, are as shown in Figure 2. Thus

$$\boldsymbol{B} = \boldsymbol{F}\boldsymbol{F}^{\mathrm{T}} = \lambda^{-1}(\boldsymbol{e}_1 \otimes \boldsymbol{e}_1 + \boldsymbol{e}_2 \otimes \boldsymbol{e}_2) + \lambda^2 \boldsymbol{e}_3 \otimes \boldsymbol{e}_3, \qquad \text{(A)}$$

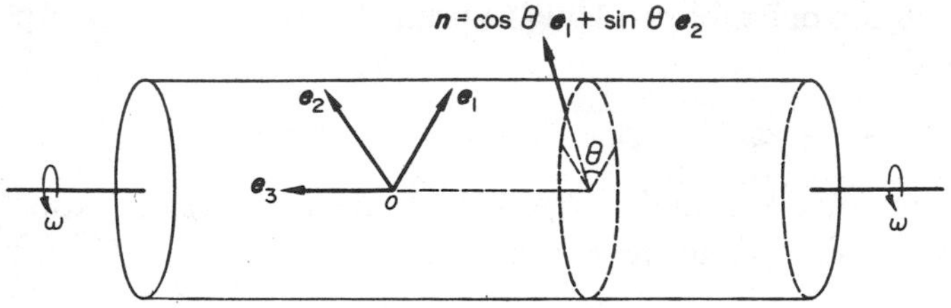

FIGURE 2 Current configuration of a circular cylinder spinning about its axis (Problem 15).

from which we see that the principal stretches (which are the square roots of the proper numbers of $\boldsymbol{B}$) are $\lambda^{-1/2}, \lambda^{-1/2}, \lambda$. The homogeneous deformation resulting from the rotation accordingly imposes on the cylinder an axial stretch λ and a transverse stretch $\lambda^{-1/2}$. Since the product of the principal stretches is unity the motion is isochoric.

Substituting the expression (A) into the Mooney constitutive equation and noting that $I_{\boldsymbol{B}} = 2\lambda^{-1} + \lambda^2$, we obtain, for the stress in the cylinder,

$$\begin{aligned}\boldsymbol{\sigma} = \{&-p + \alpha\lambda^{-1} + \beta(\lambda^{-2} + \lambda)\}\boldsymbol{I} \\ &+ (\alpha + \beta\lambda^{-1})(\lambda^2 - \lambda^{-1})\boldsymbol{e}_3 \otimes \boldsymbol{e}_3.\end{aligned} \tag{B}$$

The acceleration, found by differentiation from the given representation of the motion, is

$$\boldsymbol{a} = -\omega^2(x_1\boldsymbol{e}_1 + x_2\boldsymbol{e}_2).$$

In the absence of body forces the equation of motion (3.23) therefore yields the three scalar relations

$$\frac{\partial p}{\partial x_1} = \rho\omega^2 x_1, \quad \frac{\partial p}{\partial x_2} = \rho\omega^2 x_2, \quad \frac{\partial p}{\partial x_3} = 0,$$

which are integrable and determine the pressure to within a disposable function of t:

$$p = \tfrac{1}{2}\rho\omega^2 r^2 + \varphi(t).$$

Here $r = (x_1^2 + x_2^2)^{1/2}$ is distance from the axis of spin and ρ is the density of the rubber. The calculation of the stress vector on the curved boundary of the cylinder is carried out in the manner explained in the solution of Problem 14 and the vanishing of this vector supplies the condition

$$-p + \alpha\lambda^{-1} + \beta(\lambda^{-2} + \lambda) = 0 \quad \text{at } r = \lambda^{-1/2}A \qquad \text{(C)}$$

(the radius of the spinning cylinder being $\lambda^{-1/2}A$). Equation (C) specifies φ and leads to the expression

$$p = \tfrac{1}{2}\rho\omega^2(r^2 - \lambda^{-1}A^2) + \alpha\lambda^{-1} + \beta(\lambda^{-2} + \lambda) \qquad \text{(D)}$$

for the pressure.[15]

The stress vector on the left-hand end-face of the cylinder (as viewed in Figure 2) is

$$t_{(e_3)} = \sigma^{\mathrm{T}} e_3 = \{-\tfrac{1}{2}\rho\omega^2(r^2 - \lambda^{-1}A^2) + (\alpha + \beta\lambda^{-1})(\lambda^2 - \lambda^{-1})\}e_3,$$

use being made of (B) and (D). The stress vector on the right-hand end-face is the negative of this. The requirement of zero resultant force on the end-faces therefore takes the form

$$2\pi \int_0^{\lambda^{-1/2}A} \{-\tfrac{1}{2}\rho\omega^2(r^2 - \lambda^{-1}A^2) + (\alpha + \beta\lambda^{-1})(\lambda^2 - \lambda^{-1})\}r\,dr = 0,$$

and on performing the integration we arrive at a quartic equation for λ:

$$\alpha\lambda^4 + \beta\lambda^3 - (\alpha - \tfrac{1}{4}\rho\omega^2A^2)\lambda - \beta = 0. \qquad \text{(E)}$$

The length of the spinning cylinder is λL.

Inspection of (E) shows that there is a root between 0 and 1 and that the product of roots is negative. The possibility of there being three positive roots can be ruled out since the second derivative of the quartic would then have a positive zero which is seen not to be the case. Hence equation (E) has precisely one positive root, located in the interval (0, 1), and, as would be expected, the cylinder is made shorter and fatter by the rotation.

EXERCISES

1. The *Rivlin–Ericksen tensors* $A_1, A_2, \ldots$ are defined by

$$\overset{(n)}{C} = F^{\mathrm{T}}A_nF, \quad n = 1, 2, \ldots,$$

[15] Note that the method of finding the pressure employed here is the same as in Problem 3.7 (p. 101).

where the superscript (n) denotes the nth material derivative. Show that

$$A_1 = 2D \quad \text{and} \quad A_n = \overset{\circ}{A}_{n-1}, \quad n = 2, 3, \ldots,$$

where $\circ$ signifies the convected rate of change defined by equation $(14)_2$. Deduce that each of the Rivlin–Ericksen tensors is symmetric and objective.

2. Prove that the basic jump conditions (3.47) to (3.49) are invariant under observer transformations.

3. For each of the following constitutive equations decide whether or not the principle of objectivity is satisfied. (α and β are scalar constants, p a scalar-valued function and f a symmetric tensor-valued function.)

(i) $\sigma = -p(t)I$.

(ii) $\sigma = \alpha(F + F^{\mathrm{T}})$.

(iii) $\sigma = f(v)$.

(iv) $\sigma = \alpha\{\operatorname{grad} a + (\operatorname{grad} a)^{\mathrm{T}} + 2L^{\mathrm{T}}L\}$.

(v) $\sigma = f(b)$.

(vi) $\dot{\sigma} = W\sigma - \sigma W + (\alpha \operatorname{tr} D)I + \beta D$.

4. The stress response of a certain type of material which exhibits both elastic and viscous properties is described by the constitutive equation

$$\sigma = f(F, \dot{F}), \qquad \text{(A)}$$

f being a symmetric tensor-valued function. Investigate the restriction imposed on f by the principle of objectivity and hence show that the most general objective form of (A) is

$$\sigma = Rf(U, \dot{U})R^{\mathrm{T}}.$$

5. (*a*) A vector or tensor field is said to be *observer invariant* if it is unchanged by an observer transformation. If u is an objective vector and T an objective tensor show that $F^{\mathrm{T}}u$, $F^{-1}u$ and $F^{\mathrm{T}}TF$, $F^{-1}T(F^{-1})^{\mathrm{T}}$ are observer invariant. Deduce that the referential heat flux vector Q, the right Cauchy–Green strain tensor C and the second Piola–Kirchhoff stress tensor t are all observer invariant.

(*b*) It follows from (*a*) that the constitutive equation $\boldsymbol{t} = \boldsymbol{l}(\boldsymbol{C})$, where $\boldsymbol{l}$ is a symmetric tensor-valued function, satisfies the principle of objectivity. By using equation (22), or otherwise, show that the constitutive equation of an elastic material is necessarily of this form.

6. (i) In the text the reduced constitutive equation (34) appropriate to an isotropic elastic material is derived directly from the basic constitutive equation (4). Show that it can also be deduced from the objective form (22).

(ii) If the material considered in Exercise 4 is a fluid (i.e. if it possesses a configuration relative to which the isotropy group coincides with the unimodular group), prove that it is a Reiner–Rivlin fluid.

7. A laminated body possesses a family of material surfaces, which in some configuration are parallel planes, in which no change of area can occur. Show that, at a representative point x in the current configuration of the body, the associated constraint stress is a superposition of equal uniaxial stresses in a pair of orthogonal directions tangential to the surface of constant area through x.

8. (i) Show that, for a compressible Newtonian viscous fluid, the extra stress power $\operatorname{tr}(\boldsymbol{\sigma}^{E}\boldsymbol{D})$ is universally non-negative if and only if the bulk and shear viscosities are non-negative.

(ii) An incompressible Newtonian viscous fluid occupies a regular region B with a fixed rigid boundary ∂B and is not acted on by body forces. Show that the total kinetic energy of the fluid decreases at the rate

$$-\mu \int_B |\boldsymbol{\omega}|^2 \, dv,$$

μ being the viscosity and $\boldsymbol{\omega}$ the vorticity.

9. (*a*) Show that the circulatory motion defined in Exercise 2.9 (p. 85) can be maintained in an incompressible Newtonian viscous fluid without the aid of body forces if the function f satisfies the partial differential equation

$$\nu\left(\frac{\partial^2 f}{\partial r^2} + \frac{3}{r}\frac{\partial f}{\partial r}\right) = \frac{\partial f}{\partial t},$$

and that the pressure is then given by

$$p = p_0 + \rho \int^r \{f(s, t)\}^2 s \, ds.$$

Here ρ is the density and ν the kinematic viscosity of the fluid and p_0 is a constant.

(*b*) An incompressible Newtonian fluid is confined between long coaxial rigid cylinders of radii r_1 and r_2 ($r_1 < r_2$) which rotate about their common axis with angular speeds Ω_1 and Ω_2 respectively. Assuming that the fluid performs a steady circulatory motion and that body forces are absent, deduce from the results of part (*a*) that the circumferential velocity at distance r from the axis is

$$\frac{\Omega_2 r_2^2 - \Omega_1 r_1^2}{r_2^2 - r_1^2} r - \frac{\Omega_2 - \Omega_1}{r_2^2 - r_1^2} \frac{r_1^2 r_2^2}{r}.$$

Calculate the torque per unit length needed to drive each cylinder.

10. Two incompressible Newtonian fluids which do not mix form two layers, each of uniform thickness h, flowing steadily under gravity down a fixed plane inclined to the horizontal at an angle α. The densities of the upper and lower fluids are ρ_1 and ρ_2 respectively and the corresponding viscosities are μ_1 and μ_2. Assuming that the air above the fluids exerts a uniform pressure, show that the speed of the upper fluid at its free surface is

$$\tfrac{1}{2} g h^2 \{\rho_1(\mu_1^{-1} + 2\mu_2^{-1}) + \rho_2 \mu_2^{-1}\} \sin \alpha.$$

11. In relation to a system of rectangular Cartesian coordinates x_1, x_2, x_3, the region $-\infty < x_1 < \infty, x_2 \geqslant 0, x_3 \geqslant 0$ is filled with an incompressible Newtonian fluid. The half-planes bounding the region are occupied by porous walls through which fluid can be sucked, but relative to which it cannot slip tangentially. The fluid is in steady motion, its velocity far from the walls being in the 1-direction and of magnitude V. There is no pressure gradient in the 1-direction and no body forces act. Find the form of the function f for which the expressions

$$v_1 = Vf(x_2 x_3), \quad v_2 = -\alpha x_3, \quad v_3 = -\alpha x_2,$$

provide an exact solution for the velocity field, α being a positive constant. Comment on the character of the flow with particular reference to behaviour near the walls.

12. Show that the constitutive equation of an incompressible Reiner–Rivlin fluid can be expressed in the form

$$\boldsymbol{\sigma} = -p\boldsymbol{I} + \nu_1\boldsymbol{D} + \nu_2\boldsymbol{D}^2,$$

where the response functions ν_1 and ν_2 depend upon the principal stretching invariants $II_{\boldsymbol{D}}$ and $III_{\boldsymbol{D}}$.

Such a fluid fills the space between parallel rigid plates. One plate is held fixed and a steady shearing motion is produced in the fluid by translating the other plate in its own plane with constant speed V. Introducing a system of rectangular Cartesian coordinates x_1, x_2, x_3 in which the stationary and moving plates occupy the planes $x_2 = 0$ and $x_2 = d$ respectively and the moving plate travels in the 1-direction, and assuming that the velocity field in the fluid is of the form

$$v_1 = v(x_2), \quad v_2 = 0, \quad v_3 = 0,$$

calculate the stress components. Verify that the shear stress σ_{12} is a function (τ, say) of D_{12} only. Given that there is no pressure gradient in the 1-direction, that no body forces act, and that τ is single-valued, show that $v_1 = Vx_2/d$ and p is constant.

Specialize your results to the case in which the fluid is Newtonian. In what main respect do the two solutions differ?

13. A ball of radius A is fashioned from foam rubber of the type considered in Problem 14 (p. 156), the constant ν having the value $\frac{1}{4}$ for the particular material used. If the ball is subjected to a homogeneous compression by the application to its boundary of a uniform pressure P, show that its radius is reduced to $\gamma^{-1}A$ where

$$(\gamma^5 - 1)(\alpha\gamma + \beta) = P.$$

If $\alpha > 0$, $\beta > 0$ and $P > -\beta$, prove that this equation has precisely one positive root and that when $P > 0$ the root in question exceeds unity.

14. A rubber cylinder of radius A and length L is fixed at one end and the other end is rotated through an angle τL about the axis of symmetry. In relation to a common rectangular Cartesian system of referential and spatial coordinates with the origin at the centre of the fixed end and the 3-direction along the axis of symmetry, the resulting deformation is given by

$$x_1 = X_1 \cos \tau X_3 - X_2 \sin \tau X_3,$$
$$x_2 = X_1 \sin \tau X_3 + X_2 \cos \tau X_3, \quad x_3 = X_3.$$

Verify that this is an isochoric deformation and, assuming that the rubber is a Mooney material (conforming to the constitutive equation stated in Problem 15, p. 158), prove that the cylinder can rest permanently in its twisted configuration subject only to tractions acting on its end-faces. Show that the pressure is given by

$$p = \alpha + 2\beta + \tfrac{1}{2}\tau^2\{\alpha(A^2 - r^2) + 2\beta r^2\},$$

where $r = (x_1^2 + x_2^2)^{1/2}$, and find the resultant torque and the resultant force which must be applied to each end-face.

15. In relation to a common rectangular Cartesian system of referential and spatial coordinates a motion of a material body is given by

$$x_1 = \lambda X_1 + a \cos \gamma(X_3, t), \quad x_2 = \lambda X_2 + a \sin \gamma(X_3, t), \quad x_3 = \Lambda X_3,$$

where $\gamma(X_3, t) = (\kappa c/a)\{t - (\Lambda X_3/c)\}$ and λ, Λ, a, κ and c are positive constants. Show that this motion can be supported, without the aid of body forces, in an isotropic elastic material characterized by the constitutive equation (34) if, for given values of λ, Λ, a and κ,

$$\phi_1 + \phi_2\{\lambda^2 + (1 + \kappa^2)\Lambda^2\} > 0,$$

the arguments of the response functions ϕ_1 and ϕ_2 being

$$2\lambda^2 + (1 + \kappa^2)\Lambda^2, \quad \lambda^4 + (2 + \kappa^2)\lambda^2\Lambda^2, \quad \lambda^4\Lambda^2.$$

Describe in general terms the nature of the motion.

HINTS AND ANSWERS TO EXERCISES

Chapter 1

6. $\lambda_1 = 1, \lambda_2 = 0, \lambda_3 = -1.$

$$p_1 = \tfrac{1}{3}(-e_1 + 2e_2 + 2e_3), \quad p_2 = \tfrac{1}{3}(2e_1 - e_2 + 2e_3),$$

$$p_3 = \tfrac{1}{3}(2e_1 + 2e_2 - e_3).$$

8. (i) $W_{ij} = \mp \varepsilon_{ijp} w_p$, $w_i = \mp\tfrac{1}{2}\varepsilon_{ipq} W_{pq}$, the upper signs holding if the basis is positive in E^+ and the lower signs if it is positive in E^-.

10. If Q is an arbitrary orthogonal tensor and (e_1, e_2, e_3) is an orthonormal basis, (Qe_1, Qe_2, Qe_3) is also an orthonormal basis.

12. When $A_{23} = A_{32} = A_{31} = A_{13} = 0$ the corresponding components of U, V and Q also vanish. Explain this property and use it to simplify the calculations. In matrix form (relative to e),

$$\begin{pmatrix} 2 & 2 & 0 \\ -1 & 1 & 0 \\ 0 & 0 & -1 \end{pmatrix} = \begin{pmatrix} 1/\sqrt{2} & 1/\sqrt{2} & 0 \\ -1/\sqrt{2} & 1/\sqrt{2} & 0 \\ 0 & 0 & -1 \end{pmatrix} \begin{pmatrix} 3/\sqrt{2} & 1/\sqrt{2} & 0 \\ 1/\sqrt{2} & 3/\sqrt{2} & 0 \\ 0 & 0 & 1 \end{pmatrix}$$

$$= \begin{pmatrix} 2\sqrt{2} & 0 & 0 \\ 0 & \sqrt{2} & 0 \\ 0 & 0 & 1 \end{pmatrix} \begin{pmatrix} 1/\sqrt{2} & 1/\sqrt{2} & 0 \\ -1/\sqrt{2} & 1/\sqrt{2} & 0 \\ 0 & 0 & -1 \end{pmatrix}$$

13. A representative point is carried into its reflection in the plane through the origin orthogonal to p. Note that the given tensor is the negative of the tensor obtained by setting $\theta = \pi$ in equation (1.79). Its action is therefore equivalent to a rotation through 180° about the axis defined by p followed by the central inversion represented by $-I$.

Chapter 2

2. $\dot{\boldsymbol{r}} = \boldsymbol{W}\boldsymbol{r} = \boldsymbol{w} \wedge \boldsymbol{r}$, $\ddot{\boldsymbol{r}} = (\dot{\boldsymbol{W}} + \boldsymbol{W}^2)\boldsymbol{r} = \dot{\boldsymbol{w}} \wedge \boldsymbol{r} + \boldsymbol{w} \wedge (\boldsymbol{w} \wedge \boldsymbol{r})$.

3. $x_1 = \frac{1}{3}X_1(2 + \cos nt)$, $x_2 = \frac{1}{2}X_2(2 + \sin nt)$, $x_3 = X_3$.

Particle paths:

$$\left(\frac{x_1 - \frac{2}{3}X_1}{\frac{1}{3}X_1}\right)^2 + \left(\frac{x_2 - X_2}{\frac{1}{2}X_2}\right)^2 = 1, \quad x_3 = X_3.$$

Streamlines:

$$x_1^{2\cos nt + \cos^2 nt} x_2^{2\sin nt + \sin^2 nt} = \text{constant}, \quad x_3 = \text{constant}.$$

Examine the streamline pattern at intervals of $\pi/4n$ between $t = 0$ and $t = 2\pi/n$. $v = \frac{1}{6}(2 + \sin nt)(2 + \cos nt)V$.

4. $v_i = -\frac{1}{2}\alpha_i x_i t^{-1}$ $(i = 1, 2, 3)$. $\alpha_1 + \alpha_2 + \alpha_3 = 0$.

7. (i) All directions in the glide planes and all directions in planes which are orthogonal to the planes of shear and inclined at $2\psi = \tan^{-1}(2/\gamma)$ to the direction of shear.

All directions in planes orthogonal to the direction of shear and all directions in planes which are orthogonal to the planes of shear and inclined at $\frac{1}{2}\pi - 2\psi$ to the reversed direction of shear.

(ii) Consider in turn the cases $\lambda_1 > 1 > \lambda_2 > \lambda_3$, $\lambda_1 > \lambda_2 > 1 > \lambda_3$, $\lambda_1 > \lambda_2 = 1 > \lambda_3$ and $\lambda_1 = \lambda_2 \neq \lambda_3$ where $\lambda_1, \lambda_2, \lambda_3$ are the principal stretches.

8. (*a*) and (*b*) are unaffected. The remaining fibres are inclined at $\frac{1}{4}\pi$ to the 1-direction in the deformed configuration when the maximum contraction is imposed in the 3-direction. This statement replaces (*c*).

9. $v_1 = -v_2 = \frac{1}{2}r\,\partial f/\partial r$, $v_3 = 0$. $\boldsymbol{w} = \frac{1}{2}(2f + r\partial f/\partial r)\boldsymbol{e}_3$.

$$\boldsymbol{r}_1 = \cos(\theta + \tfrac{1}{4}\pi)\boldsymbol{e}_1 + \sin(\theta + \tfrac{1}{4}\pi)\boldsymbol{e}_2,$$

$$\boldsymbol{r}_2 = -\sin(\theta + \tfrac{1}{4}\pi)\boldsymbol{e}_1 + \cos(\theta + \tfrac{1}{4}\pi)\boldsymbol{e}_2, \quad \boldsymbol{r}_3 = \boldsymbol{e}_3,$$

where $x_1 = r\cos\theta$, $x_2 = r\sin\theta$ and $\boldsymbol{e}_1, \boldsymbol{e}_2, \boldsymbol{e}_3$ are the base vectors.

10. A direct method of proving that $\boldsymbol{A}^2$ is symmetric is via the result $\operatorname{grad} \boldsymbol{a} = \dot{\boldsymbol{L}} + \boldsymbol{L}^2$ (which should be verified).

Chapter 3

Problem 3.9 (p. 105). Since the body is permanently at rest the second law of motion $(11)_2$ gives

$$G(B; o) = \int_B \boldsymbol{x} \wedge (\rho \boldsymbol{b})\, dv + \int_{\partial B} \boldsymbol{x} \wedge \boldsymbol{t}_{(\boldsymbol{n})}\, da = \mathbf{0}.$$

Hence

$$\int_B \rho(\boldsymbol{x} \otimes \boldsymbol{b} - \boldsymbol{b} \otimes \boldsymbol{x})\, dv + \int_{\partial B} (\boldsymbol{x} \otimes \boldsymbol{t}_{(\boldsymbol{n})} - \boldsymbol{t}_{(\boldsymbol{n})} \otimes \boldsymbol{x})\, da = \mathbf{O}.$$

When this result is combined with the simplified formula (D), equation (A) is recovered.

2. Note that $\sigma = \boldsymbol{t}_{(\boldsymbol{q})} \cdot \boldsymbol{q} = \sigma_{(\boldsymbol{q})}$.

3. $\sigma_1 = \alpha + 2\beta, \quad \sigma_2 = \sigma_3 = \alpha - \beta.$

$$\boldsymbol{s}_1 = \frac{1}{\sqrt{3}}(\boldsymbol{e}_1 + \boldsymbol{e}_2 + \boldsymbol{e}_3), \quad \boldsymbol{s}_2 = \frac{1}{\sqrt{6}}(\boldsymbol{e}_1 + \boldsymbol{e}_2 - 2\boldsymbol{e}_3),$$

$$\boldsymbol{s}_3 = \frac{1}{\sqrt{2}}(\boldsymbol{e}_1 - \boldsymbol{e}_2).$$

The maximum shear stress norm is $\frac{3}{2}|\beta|$.

5. Note the relations $\sigma_i = J^{-1}\lambda_i^2 t_i$ between the principal Cauchy and Piola–Kirchhoff stresses which hold when $\boldsymbol{\sigma}$ and $\boldsymbol{B}$ (or $\boldsymbol{t}$ and $\boldsymbol{C}$) are coaxial.

6. Problem 1.20 (p. 44) can be used, in conjunction with the equation of motion (23), to put the balance law into the form

$$\int_{R_t} (\rho \boldsymbol{c} - \boldsymbol{\pi})\, dv + \int_{\partial R_t} \boldsymbol{u}_{(\boldsymbol{n})}\, da = \mathbf{0},$$

where $\boldsymbol{\pi}$ is the axial vector of $\boldsymbol{\sigma} - \boldsymbol{\sigma}^{\mathrm{T}}$. The field equation is

$$\operatorname{div} \boldsymbol{\mu} - \boldsymbol{\pi} + \rho \boldsymbol{c} = \mathbf{0}$$

and the jump condition $[\boldsymbol{u}_{(\boldsymbol{n})}] = \mathbf{0}$.

Chapter 4

2. Assume that the singular surface is given by $\Phi(\boldsymbol{x}, t) = 0$ in the frame $\mathscr{F}$ and by $\Phi'(\boldsymbol{x}', t') = 0$ in the equivalent frame $\mathscr{F}'$. Then

$$\boldsymbol{n} = |\text{grad}\, \Phi|^{-1} \text{grad}\, \Phi, \quad V_n = -|\text{grad}\, \Phi|^{-1} (\partial\Phi/\partial t),$$
$$V = -|\text{grad}\, \Phi|^{-1}\dot{\Phi},$$

with analogous results in $\mathscr{F}'$, and it follows that $V' = V$.

3. (i) Yes, provided that t is understood to mean $t - 0$ where 0 is an instant of time.

(ii) No.

(iii), (v) If and only if $\boldsymbol{f}$ is a constant spherical tensor.

(iv), (vi) Yes. Note that (iv) can be rewritten as $\boldsymbol{\sigma} = \alpha \boldsymbol{A}_2$ where $\boldsymbol{A}_2$ is the second Rivlin–Ericksen tensor (see Exercise 1 and the hint for Exercise 2.10).

7. The appropriate internal constraint is

$$\boldsymbol{N}.(\boldsymbol{C}^{-1}\boldsymbol{N}) - (\det \boldsymbol{C})^{-1} = 0$$

(see p. 63).

8. (i) Introduce the deviator of the stretching tensor.

9. (*b*) For the outer cylinder,

$$4\pi\mu r_1^2 r_2^2 \frac{\Omega_2 - \Omega_1}{r_2^2 - r_1^2}$$

in the sense of rotation, μ being the viscosity of the fluid. The torque on the inner cylinder is equal and opposite.

11. $f(x) = 1 - \exp(-\alpha x/\nu)$ where ν is the kinematic viscosity of the fluid.

There are identical boundary layers on the walls $x_2 = 0$ and $x_3 = 0$, merging near the corner. The thickness of the boundary layer on $x_2 = 0$ is of order $\nu/\alpha x_3$. Fluid from the mainstream is progressively entrained in the boundary layers. Eventually the forward speed v_1 of a typical particle falls to zero at which instant it reaches one or other of the porous walls and is absorbed.

12. There is a qualitative difference between the normal stresses given by the two solutions. In the special case of a Newtonian fluid, $-\sigma_{11}$, $-\sigma_{22}$ and $-\sigma_{33}$ are all equal to p_0, the constant ambient pressure. In the general case, however,

$$\sigma_{11} = \sigma_{22} = -p_0 + \tfrac{1}{4}(V/d)^2 \nu_2\left(-\tfrac{1}{4}(V/d)^2, 0\right), \quad \sigma_{33} = -p_0,$$

which means that the tractions which must be applied to the plates in order to maintain the shearing motion have a normal component depending upon the plate speed V.

14. Taking the basis $\{\boldsymbol{e}_1, \boldsymbol{e}_2, \boldsymbol{e}_3\}$ of the coordinate system to be positive in E^+, the resultant torque $\boldsymbol{M}$ and the resultant force $\boldsymbol{N}$ on the fixed end-face are

$$\boldsymbol{M} = -\tfrac{1}{2}\pi(\alpha + \beta)\tau A^4 \boldsymbol{e}_3 \quad \text{and} \quad \boldsymbol{N} = \tfrac{1}{4}\pi(\alpha + 2\beta)\tau^2 A^4 \boldsymbol{e}_3.$$

The resultants on the twisted end-face are $-\boldsymbol{M}$ and $-\boldsymbol{N}$.

15. The motion is a circularly polarized transverse harmonic wave of angular frequency $\kappa c/a$ and (finite) amplitude a superimposed upon a biaxial homogeneous deformation in which the principal stretches are λ, λ and Λ. The wave propagates in the 3-direction with speed c.

Appendix to Chapter 1

Section 2

The existence of the adjugate tensor, asserted between equations (37) and (38) (p. 20), may be established as follows.

Let $\boldsymbol{c}$ and $\boldsymbol{d}$ be linearly independent vectors and λ a scalar. Let $\boldsymbol{f}$ and $\boldsymbol{g}$ be vectors such that

$$\boldsymbol{f} \wedge \boldsymbol{g} = \lambda \boldsymbol{c} \wedge \boldsymbol{d}.$$

By equation (9) and axioms (6), (1), (2) and (4),

$$[\boldsymbol{f},\boldsymbol{c},\boldsymbol{d}] = 0, \quad [\boldsymbol{g},\boldsymbol{c},\boldsymbol{d}] = 0,$$

and we deduce from property (iii) of Problem 2 that $\boldsymbol{f}$ and $\boldsymbol{g}$ are expressible as linear combinations of $\boldsymbol{c}$ and $\boldsymbol{d}$:

$$\boldsymbol{f} = \alpha\boldsymbol{c} + \beta\boldsymbol{d}, \quad \boldsymbol{g} = \gamma\boldsymbol{c} + \delta\boldsymbol{d}, \text{ with } \alpha\delta - \beta\gamma = \lambda.$$

Let $\boldsymbol{A}$ be an arbitrary tensor. Then, in view of (21),

$$(\boldsymbol{Af}) \wedge (\boldsymbol{Ag}) = (\alpha\delta - \beta\gamma)(\boldsymbol{Ac}) \wedge (\boldsymbol{Ad}) = \lambda(\boldsymbol{Ac}) \wedge (\boldsymbol{Ad}).$$

The special case $\lambda = 1$ of this result shows that, for all $\boldsymbol{a}, \boldsymbol{b} \in \mathsf{E}$, $(\boldsymbol{Aa}) \wedge (\boldsymbol{Ab})$ is a function of $\boldsymbol{a} \wedge \boldsymbol{b}$. The general case affirms that the dependence of $(\boldsymbol{Aa}) \wedge (\boldsymbol{Ab})$ on $\boldsymbol{a} \wedge \boldsymbol{b}$ is linear. Thus, corresponding to $\boldsymbol{A}$ there is a unique tensor $\boldsymbol{A}^*$ (the adjugate of $\boldsymbol{A}$) such that

$$\boldsymbol{A}^*(\boldsymbol{a} \wedge \boldsymbol{b}) = (\boldsymbol{Aa}) \wedge (\boldsymbol{Ab}) \quad \forall \boldsymbol{a}, \boldsymbol{b} \in \mathsf{E}. \tag{38}$$

ADDITIONAL EXERCISES

Supplement to Exercise 4 (ii) (p. 47).

(*d*) $(\boldsymbol{A} + \alpha\boldsymbol{I})^* = \boldsymbol{A}^* - \alpha\{\boldsymbol{A}^{\mathrm{T}} - (\mathrm{tr}\boldsymbol{A})\boldsymbol{I}\} + \alpha^2\boldsymbol{I} \quad \forall\alpha \in \mathsf{R}$,

(*e*) $(\boldsymbol{a} \otimes \boldsymbol{b} + \boldsymbol{c} \otimes \boldsymbol{d})^* = (\boldsymbol{a} \wedge \boldsymbol{c}) \otimes (\boldsymbol{b} \wedge \boldsymbol{d}) \quad \forall \boldsymbol{a}, \boldsymbol{b}, \boldsymbol{c}, \boldsymbol{d} \in \mathsf{E}$.

Supplement to Exercise 7 (p. 48). Show also that $\boldsymbol{p}$ is a null vector of $\boldsymbol{S}$ (i.e. $\boldsymbol{Sp} = \boldsymbol{0}$) if $\boldsymbol{S}$ is semi-definite and $\boldsymbol{p}.(\boldsymbol{Sp}) = 0$.

Supplement to Exercise 8(i) (p. 48). Hence, or otherwise, establish the identity

$$(\boldsymbol{A} + \alpha\boldsymbol{w} \otimes \boldsymbol{w})^* = \boldsymbol{A}^* - \alpha\boldsymbol{WAW} \quad \forall\boldsymbol{A} \in \mathsf{L}, \alpha \in \mathsf{R}.$$

Supplement to Exercise 9 (p. 48).

(*c*) $\mathrm{tr}(\boldsymbol{W}_1\boldsymbol{W}_2\boldsymbol{W}_3) = -[\boldsymbol{w}_1, \boldsymbol{w}_2, \boldsymbol{w}_3]$,

$\boldsymbol{W}_3$ being a third skew-symmetric tensor and $\boldsymbol{w}_3$ its axial vector.

Supplement to Exercise 13 (p. 49). Suppose that two such tensors are applied in succession, $\boldsymbol{I} - 2\boldsymbol{p}_1 \otimes \boldsymbol{p}_1$ preceding $\boldsymbol{I} - 2\boldsymbol{p}_2 \otimes \boldsymbol{p}_2$ where $\boldsymbol{p}_1$ and $\boldsymbol{p}_2$ are distinct unit vectors. Confirm that their combined action on the positions of the points of $\mathfrak{E}$ is a rotation of amount $\varphi = 2\cos^{-1}(\boldsymbol{p}_1 . \boldsymbol{p}_2)$ about the direction defined by the unit vector $\operatorname{cosec}(\frac{1}{2}\varphi)\boldsymbol{p}_1 \wedge \boldsymbol{p}_2$.

A1. Let $\boldsymbol{a}, \boldsymbol{b}, \boldsymbol{c}, \boldsymbol{d}, \boldsymbol{e}, \boldsymbol{f}$ be arbitrary vectors. Verify that

$$\begin{vmatrix} \boldsymbol{a}.\boldsymbol{d} & \boldsymbol{a}.\boldsymbol{e} & \boldsymbol{a}.\boldsymbol{f} \\ \boldsymbol{b}.\boldsymbol{d} & \boldsymbol{b}.\boldsymbol{e} & \boldsymbol{b}.\boldsymbol{f} \\ \boldsymbol{c}.\boldsymbol{d} & \boldsymbol{c}.\boldsymbol{e} & \boldsymbol{c}.\boldsymbol{f} \end{vmatrix} = [\boldsymbol{a}, \boldsymbol{b}, \boldsymbol{c}][\boldsymbol{d}, \boldsymbol{e}, \boldsymbol{f}].$$

Hence, or otherwise, show that

$$\det(\boldsymbol{a} \otimes \boldsymbol{b} + \boldsymbol{c} \otimes \boldsymbol{d} + \boldsymbol{e} \otimes \boldsymbol{f}) = [\boldsymbol{a}, \boldsymbol{c}, \boldsymbol{e}][\boldsymbol{b}, \boldsymbol{d}, \boldsymbol{f}].$$

A2. Let $\boldsymbol{S}$ be an arbitrary symmetric tensor, $\boldsymbol{n}$ an arbitrary unit vector and α an arbitrary scalar. Show that the principal invariants of the tensor $\hat{\boldsymbol{S}} = \boldsymbol{S} + \alpha \boldsymbol{n} \otimes \boldsymbol{n}$ are

$$I_{\hat{S}} = I_S + \alpha, \quad II_{\hat{S}} = II_S + \alpha\{\mathrm{tr}\boldsymbol{S} - \boldsymbol{n}.(\boldsymbol{S}\boldsymbol{n})\}, \quad III_{\hat{S}} = III_S + \alpha \boldsymbol{n}.(\boldsymbol{S}^* \boldsymbol{n}).$$

If $\boldsymbol{S}$ has proper numbers $\lambda_1 < \lambda_2 < \lambda_3$, no proper vector of $\boldsymbol{S}$ is orthogonal to $\boldsymbol{n}$ and $\alpha > 0$, show that the proper numbers $\hat{\lambda}_1, \hat{\lambda}_2, \hat{\lambda}_3$ of $\hat{\boldsymbol{S}}$, appropriately ordered, have the interleaving property

$$\lambda_1 < \hat{\lambda}_1 < \lambda_2 < \hat{\lambda}_2 < \lambda_3 < \hat{\lambda}_3.$$

How is this result changed when $\alpha < 0$?

A3. Let $\boldsymbol{A}$ be an arbitrary invertible tensor and let $\boldsymbol{S} = \boldsymbol{A}^{\mathrm{T}}\boldsymbol{A}$ (which, by Problem 11, p. 27, is positive definite symmetric) have proper numbers $\lambda_1 < \lambda_2 < \lambda_3$ and associated orthonormal proper vectors $\boldsymbol{p}_1, \boldsymbol{p}_2, \boldsymbol{p}_3$. Show that

$$\boldsymbol{S} = \lambda_2 \boldsymbol{I} - \tfrac{1}{2}(\lambda_3 - \lambda_1)(\boldsymbol{q} \otimes \boldsymbol{r} + \boldsymbol{r} \otimes \boldsymbol{q}),$$

where

$$(\lambda_3 - \lambda_1)^{1/2}\boldsymbol{q} = (\lambda_2 - \lambda_1)^{1/2}\boldsymbol{p}_1 + (\lambda_3 - \lambda_2)^{1/2}\boldsymbol{p}_3,$$

$$(\lambda_3 - \lambda_1)^{1/2}\boldsymbol{r} = (\lambda_2 - \lambda_1)^{1/2}\boldsymbol{p}_1 - (\lambda_3 - \lambda_2)^{1/2}\boldsymbol{p}_3.$$

A unit vector $\boldsymbol{n}$ is said to define *a circle $C_{(\boldsymbol{n})}$ of special directions in relation to* $\boldsymbol{A}$ if $|\boldsymbol{A}\boldsymbol{m}|$ has the same value for all unit vectors $\boldsymbol{m}$ orthogonal to $\boldsymbol{n}$. Verify that $C_{(\boldsymbol{q})}$ and $C_{(\boldsymbol{r})}$ are such circles and prove that there are no others. If $\boldsymbol{m}_1$ and $\boldsymbol{m}_2$ are unit vectors defining special directions in either $C_{(\boldsymbol{q})}$ or $C_{(\boldsymbol{r})}$, show that the angles between $\boldsymbol{m}_1, \boldsymbol{m}_2$ and $\boldsymbol{A}\boldsymbol{m}_1, \boldsymbol{A}\boldsymbol{m}_2$ are equal.

A4. If $\boldsymbol{l}, \boldsymbol{m}, \boldsymbol{n}$ are orthonormal vectors show that

$$\boldsymbol{Q} = \boldsymbol{m} \otimes \boldsymbol{n} + \boldsymbol{n} \otimes \boldsymbol{l} + \boldsymbol{l} \otimes \boldsymbol{m}$$

is an orthogonal tensor. Show further that $\boldsymbol{Q}^2 = \boldsymbol{Q}^{\mathrm{T}}$ and deduce that $\boldsymbol{Q}$ is proper orthogonal.

A5. Verify that the representation (79) (p. 32) of a proper orthogonal tensor can be put into the form

$$\boldsymbol{Q} = \boldsymbol{I} + \sin\theta \boldsymbol{P} + (1 - \cos\theta)\boldsymbol{P}^2,$$

where $\boldsymbol{P}$ is the skew-symmetric tensor having as its axial vector the unit vector $\boldsymbol{p}$ such that $\boldsymbol{Qp} = \boldsymbol{p}$. Derive the relations

$$\boldsymbol{P}^{2n} = (-1)^{n+1}\boldsymbol{P}^2, \quad \boldsymbol{P}^{2n+1} = (-1)^n \boldsymbol{P}, \quad n = 1, 2, \ldots,$$

and hence show that

$$\boldsymbol{Q} = \exp(\theta \boldsymbol{P}).$$

Deduce that successive rotations through angles θ_1 and θ_2 about the same axis are equivalent to a single rotation about that axis of amount $\theta_1 + \theta_2$.

HINTS

A1. The first result leads easily to the formula for $\varepsilon_{ijk}\varepsilon_{lmn}$ in Exercise 3 (p. 47).

A2. Let $\boldsymbol{p}_1, \boldsymbol{p}_2, \boldsymbol{p}_3$ be orthonormal proper vectors of $\boldsymbol{S}$ corresponding to λ_1, λ_2, λ_3. Show that the characteristic equation of $\hat{\boldsymbol{S}}$ is

$$(\lambda - \lambda_1)(\lambda - \lambda_2)(\lambda - \lambda_3)$$
$$-\alpha\{n_1^2(\lambda - \lambda_2)(\lambda - \lambda_3) + n_2^2(\lambda - \lambda_3)(\lambda - \lambda_1) + n_3^2(\lambda - \lambda_1)(\lambda - \lambda_2)\} = 0,$$

where $n_i = \boldsymbol{n}.\boldsymbol{p}_i$ $(i = 1, 2, 3)$. Examine the sign of the left-hand side at $\lambda = \lambda_i$.

A3. A circle of special directions exists if there are unit vectors $\boldsymbol{l}, \boldsymbol{m}$, forming an orthonormal set with $\boldsymbol{n}$, such that $|\boldsymbol{A}(\cos\theta\, \boldsymbol{l} + \sin\theta\, \boldsymbol{m})|$ is independent of θ. Show that $\boldsymbol{l}$ and $\boldsymbol{m}$ are then orthogonal to either $\boldsymbol{q}$ or $\boldsymbol{r}$.

A5. The first result can be used to solve Exercise 11 (p. 49). If $\boldsymbol{T}$ is a tensor which commutes with every proper orthogonal tensor, first prove that every unit vector is a proper vector of $\boldsymbol{T}$. Then, by considering the action of $\boldsymbol{T}$ on $3^{-1/2}(\boldsymbol{l} + \boldsymbol{m} + \boldsymbol{n})$, where $\boldsymbol{l}, \boldsymbol{m}, \boldsymbol{n}$ are orthonormal, show that the proper numbers of $\boldsymbol{T}$ corresponding to $\boldsymbol{l}, \boldsymbol{m}, \boldsymbol{n}$ are equal.

Appendix to Chapter 2

ADDITIONAL EXERCISES

A1. A motion of a body is given by

$$x_1 = e^{-at}(X_1 \cos bX_3 t - X_2 \sin bX_3 t), \quad x_2 = e^{-at}(X_1 \sin bX_3 t + X_2 \cos bX_3 t), \quad x_3 = \phi(t)X_3,$$

where a and b are positive constants. Find the form of $\phi(t)$ for which the motion is isochoric. Determine, for the isochoric motion, the components of velocity and acceleration in the spatial description and show that both the particle paths and the streamlines lie on the surfaces $(x_1^2 + x_2^2)x_3 =$ constant.

A2. A plane circular shearing motion of a body is given by

$$x_1 = X_1 + \phi(X_3)\cos\omega t + \psi(X_3)\sin\omega t, \quad x_2 = X_2 + \phi(X_3)\sin\omega t - \psi(X_3)\cos\omega t, \quad x_3 = X_3,$$

where ω is a positive constant, the functions ϕ and ψ are differentiable and the referential and spatial coordinates refer to a common rectangular Cartesian system. Show that this motion is isochoric and that the particle paths and streamlines are circles. Discuss the extension and rotation undergone by material line elements which, in the reference configuration, lie parallel and orthogonal to the 3-direction.

A3. Show that the deformation gradient $\boldsymbol{F}$ has the representation

$$\boldsymbol{F} = \sum_{r=1}^{3} \lambda_r (\boldsymbol{q}_r \otimes \boldsymbol{p}_r),$$

where λ_i $(i = 1,2,3)$ are the principal stretches and the orthonormal vectors $\boldsymbol{p}_i$ and $\boldsymbol{q}_i$ are directed along the referential and current stretch axes, respectively. Deduce the formula

$$\boldsymbol{L} = \sum_{r=1}^{3} \frac{\dot{\lambda}_r}{\lambda_r} \boldsymbol{q}_r \otimes \boldsymbol{q}_r - \boldsymbol{F}\boldsymbol{M}\boldsymbol{F}^{-1} + \boldsymbol{N}$$

for the velocity gradient where

$$\boldsymbol{M} = \dot{\boldsymbol{p}}_r \otimes \boldsymbol{p}_r \text{ and } \boldsymbol{N} = \dot{\boldsymbol{q}}_r \otimes \boldsymbol{q}_r$$

are skew-symmetric tensors representing the spins of the referential and current stretch axes. Hence prove that

(*a*) if the referential stretch axes are fixed the current stretch axes are principal axes of stretching,

(*b*) if the current stretch axes are fixed they are principal axes of the strain rate $\dot{\boldsymbol{B}}$,

(*c*) if both the referential and current stretch axes are fixed the spin tensor $\boldsymbol{W}$ is zero.

A4. A vector field $\boldsymbol{H}$, defined on the configurations of a moving body, satisfies the equations

$$\frac{\partial \boldsymbol{H}}{\partial t} = \operatorname{curl}(\boldsymbol{v} \wedge \boldsymbol{H}), \quad \operatorname{div} \boldsymbol{H} = 0,$$

$\boldsymbol{v}$ being the velocity. Show that

$$\frac{d}{dt}\int_{S_t} \boldsymbol{H}.\boldsymbol{n}\mathrm{d}a = 0,$$

where S_t is the current configuration of an arbitrary material surface. Show also that the field lines of $\boldsymbol{H}$ are material curves.

A5. Γ is a circuit, Π a plane and $\boldsymbol{l}$ a unit vector normal to Π. Verify that the area enclosed by the projection of Γ on Π is

$$\tfrac{1}{2}\oint_{\Gamma}(\boldsymbol{l} \wedge \boldsymbol{x}).\mathrm{d}\boldsymbol{x}.$$

A motion of a body has the property that, for every material circuit, the area enclosed by the projection of the circuit on the plane $x_3 = 0$ does not vary with time. If Γ_t is the current configuration of an arbitrary material circuit show that

$$\oint_{\Gamma_t}(\boldsymbol{e}_3 \wedge \boldsymbol{v}).\mathrm{d}\boldsymbol{x} = 0.$$

Deduce that the velocity components $v_1 = \boldsymbol{v}.\boldsymbol{e}_1$ and $v_2 = \boldsymbol{v}.\boldsymbol{e}_2$ are independent of x_3. If the motion is isochoric what more can be said about the velocity field?

Finally, show that a motion is a rigid translation if and only if, for every material circuit, the area enclosed by the projection of the circuit on *any* plane is invariable.

HINTS AND ANSWERS

A1. $\phi(t) = e^{2at}$.

$$v_1 = -ax_1 - be^{-2at}x_2x_3, \quad v_2 = -ax_2 + be^{-2at}x_1x_3, \quad v_3 = 2ax_3.$$

$$a_1 = a^2x_1 + 2abe^{-2at}x_2x_3 - b^2e^{-4at}x_1x_3^2,$$

$$a_2 = a^2x_2 - 2abe^{-2at}x_1x_3 - b^2e^{-4at}x_2x_3^2,$$

$$a_3 = 4a^2x_3.$$

Show that the streamlines satisfy

$$2(x_1\mathrm{d}x_1 + x_2\mathrm{d}x_2)x_3 + (x_1^2 + x_2^2)\mathrm{d}x_3 = 0.$$

A2. A material line element which is parallel to the 3-direction in the reference configuration is extended by an amount $\varsigma^{1/2}$ and rotated through an angle $\cos^{-1}\varsigma^{-1/2}$ where

$$\varsigma = 1 + \{\phi'(X_3)\}^2 + \{\psi'(X_3)\}^2.$$

A material line element which is orthogonal to the 3-direction in the reference configuration is neither extended nor rotated.

A4. Observe that $\boldsymbol{H}$ behaves like the vorticity in a circulation preserving motion. Problem 14 (p. 80) can therefore be applied to the field lines of $\boldsymbol{H}$. By virtue of the properties established in this exercise $\boldsymbol{H}$ is said to be a *frozen in field.*

A5. If the motion is isochoric the entire velocity field is independent of x_3.

Appendix to Chapter 3

Section 3

Details of the derivation of equation (15) (p. 95), excluded from the original edition by space limitations, are as follows.

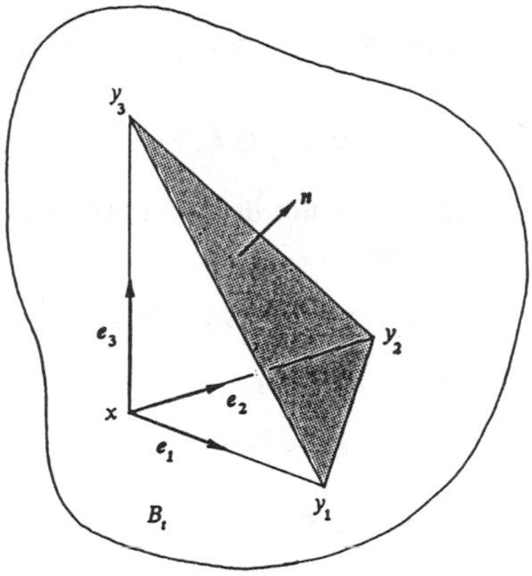

Tetrahedron construction used in deriving equation (15).

The central result of the theory of stress, specifying the dependence of the stress vector $\boldsymbol{t}_{(\boldsymbol{n})}$ on the unit normal $\boldsymbol{n}$, is obtained by applying equation $(13)_1$ to the tetrahedral subregion of B_t illustrated in the figure. The vertex x is a representative interior point of B_t and the edges xy_1, xy_2, xy_3 are aligned with the members of an arbitrarily chosen orthonormal basis e = $\{\boldsymbol{e}_1, \boldsymbol{e}_2, \boldsymbol{e}_3\}$ as shown. The distance from x to the slant face $y_1y_2y_3$ is denoted by d and the outward unit normal to $y_1y_2y_3$ by $\boldsymbol{n}$. The length of the edge xy_i $(i = 1,2,3)$ is hence $n_i^{-1}d$, where $n_i = \boldsymbol{n}.\boldsymbol{e}_i$, and we note that $n_i > 0$ by construction. The outward unit normal to the face orthogonal to xy_i is $-\boldsymbol{e}_i$.

With these facts in mind we deduce from equation $(13)_1$ (p. 94) that the balance of linear momentum for the material currently occupying the region $x\,y_1y_2y_3$ is

$$\tfrac{1}{6}(n_1n_2n_3)^{-1}d^3\bar{\rho}(\bar{\boldsymbol{a}} - \bar{\boldsymbol{b}}) = \tfrac{1}{2}(n_1n_2n_3)^{-1}d^2\{\bar{\boldsymbol{t}}_{(\boldsymbol{n})} + n_p\bar{\boldsymbol{t}}_{(-\boldsymbol{e}_p)}\}. \tag{A}$$

In each term of (A) use has been made of the mean value theorem for multiple integrals (see, for, example, T.M. Apostol, *Mathematical Analysis* (Reading, Mass. etc., Addison-Wesley, 1963), p. 269), an overbar indicating evaluation at some point of the region or surface in question. On cancelling the common factor $\tfrac{1}{2}(n_1n_2n_3)^{-1}d^2$ in (A), then making the tetrahedron shrink up to x by letting d tend to zero we arrive at the relation

$$\boldsymbol{t}_{(\boldsymbol{n})} = -(\boldsymbol{n}.\boldsymbol{e}_p)\boldsymbol{t}_{(-\boldsymbol{e}_p)}\,, \tag{B}$$

the stress vectors now being evaluated at x. Equation (A) is meaningful only when the components n_i are non-zero, but, because of the continuous dependence of $\boldsymbol{t}_{(\boldsymbol{n})}$ on $\boldsymbol{n}$, (B) remains valid when this restriction is removed. We can therefore set $\boldsymbol{n} = \boldsymbol{e}_i$ in (B), obtaining

$$\boldsymbol{t}_{(\boldsymbol{e}_i)} = -\boldsymbol{t}_{(-\boldsymbol{e}_i)}.$$

It follows that (B) can be replaced by

$$\boldsymbol{t}_{(\boldsymbol{n})} = (\boldsymbol{n}.\boldsymbol{e}_p)\boldsymbol{t}_{(\boldsymbol{e}_p)} = \{\boldsymbol{t}_{(\boldsymbol{e}_p)} \otimes \boldsymbol{e}_p\}\boldsymbol{n}. \tag{C}$$

The tensor

$$\boldsymbol{\sigma} = \boldsymbol{e}_p \otimes \boldsymbol{t}_{(\boldsymbol{e}_p)} \tag{D}$$

is plainly independent of $\boldsymbol{n}$ and it does not depend either on the choice of basis. For if $e' = \{e_1', e_2', e_3'\}$ is a second orthonormal basis,

$$\boldsymbol{\sigma} = (\boldsymbol{e}_p' \otimes \boldsymbol{e}_p')\boldsymbol{\sigma} = \boldsymbol{e}_p' \otimes (\boldsymbol{\sigma}^{\mathrm{T}}\boldsymbol{e}_p') = \boldsymbol{e}_p' \otimes \boldsymbol{t}_{(\boldsymbol{e}_p')},$$

use being made of (C) and (D). Equation (C), rewritten as

$$\boldsymbol{t}_{(\boldsymbol{n})} = \boldsymbol{\sigma}^{\mathrm{T}}\boldsymbol{n}, \tag{15}$$

thus makes explicit the dependence of $\boldsymbol{t}_{(\boldsymbol{n})}$ on $\boldsymbol{n}$. $\boldsymbol{\sigma}$ evidently provides a tensorial description of the contact forces acting at x and is accordingly called the *stress tensor*.

Section 4

An entirely algebraic solution to Problem 8 (p. 103) can be given as follows.
First replace equation (C) by

$$\tau_{(\boldsymbol{n})} = (\delta_1^2 n_2^2 n_3^2 + \delta_2^2 n_3^2 n_1^2 + \delta_3^2 n_1^2 n_2^2)^{1/2}, \tag{C}$$

with

$$\delta_1 = \sigma_2 - \sigma_3, \quad \delta_2 = \sigma_3 - \sigma_1, \quad \delta_3 = \sigma_1 - \sigma_2. \tag{D}$$

(*a*) Suppose that the principal stresses are distinct and are so labelled that $\sigma_1 < \sigma_2 < \sigma_3$. From (A) and (C),

$$\sigma_3 - \sigma_{(\boldsymbol{n})} = \delta_2 n_1^2 - \delta_1 n_2^2 \geq 0, \quad \sigma_{(\boldsymbol{n})} - \sigma_1 = \delta_2 n_3^2 - \delta_3 n_2^2 \geq 0,$$

proving that $\sigma_1 \leq \sigma_{(\boldsymbol{n})} \leq \sigma_3$. The upper and lower bounds are reached if and only if $n_3^2 = 1$ and $n_1^2 = 1$, respectively. Thus the maximum value of $\sigma_{(\boldsymbol{n})}$ at x is σ_3, attained when the normal direction $\boldsymbol{n}$ is aligned with the principal axis corresponding to σ_3. The minimum value is σ_1, again realised when $\boldsymbol{n}$ is aligned with the corresponding principal axis. At both these extrema we see from (C) that $\tau_{(\boldsymbol{n})} = 0$. This, the minimum value of the shear stress norm, is also achieved when $\boldsymbol{n} = \pm \boldsymbol{s}_2$.

With the use of the relations $\delta_2 = \delta_3 - \delta_1$ (provided by (D)) and $n_1^2 + n_2^2 + n_3^2 = 1$, the right-hand side of equation (C) can be manipulated as follows

$$\tau_{(\boldsymbol{n})} = [\delta_3^2\{\tfrac{1}{4} - (\tfrac{1}{2} - n_1^2)^2\} + \delta_1^2\{\tfrac{1}{4} - (\tfrac{1}{2} - n_3^2)^2\} + 2\delta_3\delta_1 n_3^2 n_1^2]^{1/2}$$

$$= [\tfrac{1}{4}\delta_2^2 - \{\delta_3(\tfrac{1}{2} - n_1^2) - \delta_1(\tfrac{1}{2} - n_3^2)\}^2 - \delta_3\delta_1 n_2^2]^{1/2}. \tag{E}$$

Inspection of (E) shows that $\tau_{(n)}$ takes its maximum value, $\frac{1}{2}\delta_2$, when

$$\delta_3(\tfrac{1}{2}-n_1^2)-\delta_1(\tfrac{1}{2}-n_3^2)=0 \text{ and } n_2=0. \tag{F}$$

In view of $(\mathrm{F})_2$, $\frac{1}{2}-n_1^2=-(\frac{1}{2}-n_3^2)$, whence, from $(\mathrm{F})_1$, $n_1^2=n_3^2=\frac{1}{2}$. The maximum value of $\tau_{(n)}$ at x is therefore half the difference between the greatest and least principal stresses. The corresponding normal directions are orthogonal to the principal axis associated with the intermediate principal stress σ_2 and inclined at either $\frac{1}{4}\pi$ or $\frac{3}{4}\pi$ to the other principal axes.

(*b*) Considering next the case in which two of the principal stresses are equal, suppose that $\sigma_1=\sigma_2<\sigma_3$. Then $\delta_1=-\delta_2$, $\delta_3=0$. As in case (*a*) the maximum value of $\sigma_{(n)}$ at x is σ_3, attained when $\boldsymbol{n}=\pm\boldsymbol{s}_3$. The minimum value is σ_1, reached when $n_3=0$, i.e. when $\boldsymbol{n}$ is orthogonal to $\boldsymbol{s}_3$. For $\boldsymbol{n}=\pm\boldsymbol{s}_3$ and for all $\boldsymbol{n}$ orthogonal to $\boldsymbol{s}_3$ it follows from (C) that $\tau_{(n)}=0$. We see from (E) that the maximum value of $\tau_{(n)}$ is $\frac{1}{2}(\sigma_3-\sigma_1)$, achieved now when $n_3^2=\frac{1}{2}$, that is when the normal direction lies in a circular cone of half-angle $\frac{1}{4}\pi$ about the principal axis of stress associated with σ_3. A similar state of affairs holds when $\sigma_1<\sigma_2=\sigma_3$.

(*c*) Finally, if $\sigma_1=\sigma_2=\sigma_3$ the stress at x is spherical and $\sigma_{(n)}=\sigma_1$, $\tau_{(n)}=0$ for all orientations of $\boldsymbol{n}$.

Section 5

The details of the derivation of the heat flux relation (36) (p. 111) are directly analogous to those given above in connection with the stress relation (15). A similar remark applies to the 'tetrahedron argument' required in Exercise 6 (p. 122).

ADDITIONAL EXERCISES

A1. The stress tensor at a point x has components σ_{ij} relative to an orthonormal basis $\{\boldsymbol{e}_1, \boldsymbol{e}_2, \boldsymbol{e}_3\}$, $\boldsymbol{e}_3$ being directed along a principal axis of stress at x. Verify that σ_{33} is a principal stress and that $\sigma_{31}=\sigma_{23}=0$. Show that the other principal stresses, σ_1 and σ_2, are given by the following geometrical construction. In relation to axes Oxy, draw the line segment L joining the points with coordinates $(\sigma_{11}, \sigma_{12})$ and $(\sigma_{22}, -\sigma_{12})$, and the circle M having L as a diameter. Then the points of intersection of M with Ox have coordinates $(\sigma_1, 0)$ and $(\sigma_2, 0)$. Show further that if σ is the normal stress and τ the shear stress norm at x on a surface segment with normal orthogonal to $\boldsymbol{e}_3$, the point with coordinates (σ, τ) lies on M. [M is called *Mohr's circle*.]

A2. A material body is in equilibrium and, in relation to a rectangular Cartesian system of spatial coordinates x_1, x_2, x_3, the stress $\boldsymbol{\sigma}$ does not vary with x_3. Assuming that no body forces act, prove that there exists a vector function $\boldsymbol{\phi}$ such that

$$\sigma_{1i}=-\frac{\partial\phi_i}{\partial x_2}, \quad \sigma_{2i}=\frac{\partial\phi_i}{\partial x_1}.$$

The current configuration of the body is a cylinder with its generators parallel to the x_3-axis and the boundary of the region in which the cylinder intersects a plane x_3 = constant is a circuit C. Show that the stress vector on C is $-\partial\boldsymbol{\phi}/\partial s$ where s is arc length on C measured in the sense which rotates the x_1- into the x_2-axis.

A3. The state of stress in the material body considered in Problem 8 (p. 64) is said to be *spherically symmetric* if the principal axes of stress are defined by the unit vectors $\hat{\boldsymbol{r}}$, $\hat{\boldsymbol{\theta}}$, $\hat{\boldsymbol{\phi}}$ and the associated principal stresses σ_r, σ_θ, σ_ϕ are functions of r and t only satisfying the relation $\sigma_\theta = \sigma_\phi$. Show that

$$\boldsymbol{\sigma} = \sigma_\theta \boldsymbol{I} + r^{-2}(\sigma_r - \sigma_\theta)\boldsymbol{x}\otimes\boldsymbol{x}.$$

Remembering that the velocity is purely radial, i.e. $\boldsymbol{v} = v\hat{\boldsymbol{r}}$, deduce that, in the absence of body forces, the equation of motion reduces to the single scalar relation

$$\rho\left(\frac{\partial v}{\partial t} + v\frac{\partial v}{\partial r}\right) = \frac{\partial \sigma_r}{\partial r} + \frac{2}{r}(\sigma_r - \sigma_\theta).$$

Suppose now that the body is composed of an incompressible material and that at time t it occupies the region $a(t) \le r \le b(t)$. The motion is produced by applying a uniform pressure $P(t)$ to the internal boundary, the external boundary being traction-free. Show that

$$P = \rho\left(1 - \frac{a}{b}\right)\left\{a\ddot{a} + \tfrac{1}{2}\dot{a}^2\left(1 - \frac{a}{b}\right)\left(3 + 2\frac{a}{b} + \frac{a^2}{b^2}\right)\right\} - 2\int_a^b (\sigma_r - \sigma_\theta)\frac{dr}{r}.$$

A4. An incompressible ideal fluid of density ρ is contained in a fixed right circular cylinder. The fluid is acted on by a body force

$$\boldsymbol{b} = (cx_1 + dx_2)\boldsymbol{e}_1 + (ex_1 + fx_2)\boldsymbol{e}_2$$

per unit mass where the spatial coordinates x_1, x_2, x_3 relate to an orthonormal basis $\{\boldsymbol{e}_1, \boldsymbol{e}_2, \boldsymbol{e}_3\}$, with $\boldsymbol{e}_3$ directed along the axis of the cylinder, and c, d, e, f are functions of the time t only. Show that a possible motion of the fluid is a rigid body rotation about the axis of the cylinder with angular speed Ω satisfying $d\Omega/dt = \frac{1}{2}(e - d)$ Verify that, in this motion, the pressure in the fluid is given by

$$p - p_0 = \tfrac{1}{2}\rho\{(\Omega^2 + c)x_1^2 + (d + e)x_1x_2 + (\Omega^2 + f)x_2^2\},$$

p_0 being the pressure on the axis.

A5. Derive from the basic jump conditions (47) to (49) (p. 118) the relations

(*a*) $(\rho V)^+[\tfrac{1}{2}\boldsymbol{v}.\boldsymbol{v} + \varepsilon] = [\boldsymbol{j}].\boldsymbol{n}$,

(*b*) $(\rho V)^+[\tfrac{1}{2}\boldsymbol{v}.\boldsymbol{v} - \varepsilon] = (\rho V\boldsymbol{v} + \boldsymbol{\sigma}^{\mathrm{T}}\boldsymbol{n})^+.[\boldsymbol{v}] + [\boldsymbol{q}].\boldsymbol{n}$,

where $\boldsymbol{\sigma}$ is the stress, $\boldsymbol{q}$ the heat flux vector and $\boldsymbol{j} = -\boldsymbol{\sigma v} - \boldsymbol{q}$ the energy flux vector.

Deduce from (*b*) that, in a non-heat-conducting body (for which q is identically zero), the jumps in the kinetic and internal energies per unit mass on a shock wave advancing into stationary, stress-free material are equal.

HINT

A2. An explicit formula for the components of the stress vector is

$$\phi_i(x_1, x_2) = \int_{a_1}^{x_1} \sigma_{2i}(\xi, a_2)d\xi - \int_{a_2}^{x_2} \sigma_{1i}(x_1, \eta)d\eta,$$

where, for all relevant values of x_3, the point with coordinates (a_1, a_2, x_3) is in the equilibrium configuration of the body.

Appendix to Chapter 4

ADDITIONAL EXERCISES

A1. Deduce from the energy equation (39) (p. 112) that, in the absence of heat supply, the steady flow of heat in a rigid body is governed by the equation

$$\operatorname{div} \boldsymbol{q} = 0.$$

Such a body occupies a region B with boundary ∂B, and segments S_1 and S_2 of ∂B are maintained at uniform temperatures θ_1 and θ_2, respectively. The remainder of ∂B is thermally insulated, so that $\boldsymbol{q}.\boldsymbol{n} = 0$ on this portion, $\boldsymbol{n}$ being the outward unit normal. Obtain the relation

$$\int_B \boldsymbol{q}\cdot \operatorname{grad}\theta \, dv = (\theta_1 - \theta_2)\int_{S_1} \boldsymbol{q}.\boldsymbol{n} da.$$

Hence show that if heat flows into B over S_1 and out over S_2 the *heat conduction inequality*

$$\int_B \boldsymbol{q}\cdot \operatorname{grad}\theta \, dv \geq 0$$

is a necessary and sufficient condition for S_2 not to be hotter than S_1.

A2. Show that the *Jaumann stress rate* $\overset{\vee}{\sigma}$, defined by

$$\overset{\vee}{\sigma} = \dot{\sigma} - \boldsymbol{W}\sigma + \sigma\boldsymbol{W},$$

is an objective tensor. Deduce that the constitutive equation

$$\overset{\vee}{\sigma} = \alpha(\operatorname{tr}\boldsymbol{D})\boldsymbol{I} + \beta\boldsymbol{D}, \tag{A}$$

in which α and β are constant scalars, satisfies the principle of objectivity.

A body made of the *hypoelastic* material characterized by (A) is subjected to the simple shearing motion specified in Problem 12, p. 76. If the stress in the body is uniform at all times and zero at $t = 0$, show that

$$\sigma = \tfrac{1}{2}\beta\{(1 - \cos ct)(\boldsymbol{p}\otimes\boldsymbol{p} - \boldsymbol{q}\otimes\boldsymbol{q}) + \sin ct(\boldsymbol{p}\otimes\boldsymbol{q} + \boldsymbol{q}\otimes\boldsymbol{p})\}.$$

Does this behaviour seem to you physically realistic?

A3. A material body is subjected to restricted shear in the following sense: at a representative point x in the current configuration the angle between intersecting material curves which, in the reference configuration, are tangential at the point of intersection to fixed unit vectors $\boldsymbol{L}$ and $\boldsymbol{M}$ is $\cos^{-1}(\boldsymbol{L}.\boldsymbol{M})$. Show that the associated constraint stress is a scalar multiple of

$$\boldsymbol{l}\otimes\boldsymbol{m} + \boldsymbol{m}\otimes\boldsymbol{l} - (\boldsymbol{l}.\boldsymbol{m})(\boldsymbol{l}\otimes\boldsymbol{l} + \boldsymbol{m}\otimes\boldsymbol{m}),$$

where $\boldsymbol{l}$ and $\boldsymbol{m}$ are the unit vectors tangential at x to the intersecting curves in the current configuration.

A4. If, in Problem 11, p. 151, the fluid is confined to the region $0 \le x_2 \le h$ and no shear stress acts on the surface $x_2 = h$, find the velocity field. Show that the oscillatory motion of the surface $x_2 = h$ has amplitude

$$V\{\tfrac{1}{2}(\cos 2kh + \cosh 2kh)\}^{-1/2}$$

and lags behind the vibration of the plate by the phase angle $\tan^{-1}(\tan kh \tanh kh)$.

A5. Show that in any deformation of an isotropic elastic body, made of unconstrained or incompressible material,

$$\boldsymbol{\sigma B} = \boldsymbol{B \sigma} \tag{A}$$

at each point. [(A) is the *general universal relation.*] If

$$\boldsymbol{B} = B_{11}\boldsymbol{e}_1 \otimes \boldsymbol{e}_1 + B_{22}\boldsymbol{e}_2 \otimes \boldsymbol{e}_2 + B_{33}\boldsymbol{e}_3 \otimes \boldsymbol{e}_3 + B_{12}(\boldsymbol{e}_1 \otimes \boldsymbol{e}_2 + \boldsymbol{e}_2 \otimes \boldsymbol{e}_1),$$

where $\{\boldsymbol{e}_1, \boldsymbol{e}_2, \boldsymbol{e}_3\}$ is an orthonormal basis, confirm that (A) yields the single relation

$$\sigma_{(\boldsymbol{e}_1)} - \sigma_{(\boldsymbol{e}_2)} = \{(B_{11} - B_{22})/B_{12}\}\boldsymbol{e}_1 \cdot (\boldsymbol{\sigma e}_2).$$

Recover the universal relation for simple shear obtained in Problem 13 (p. 155).

HINTS AND ANSWERS

A2. The stress is zero at $t = 0$ and at intervals of $2\pi/c$/ thereafter. It is difficult to envisage an actual material behaving in this way in a simple shearing motion. Note that proving the objectivity of (A) is equivalent to part (vi) of Exercise 3 (p. 161).

A3. The appropriate internal constraint is given by part (ii) of Problem 7 (pp. 62, 63).

A4. $v_1 = v(x_2, t)$, $v_2 = 0$, $v_3 = 0$, where

$$\begin{aligned} v(x_3, t) = {} & V(\cos 2kh + \cosh 2kh)^{-1} \\ & \times[\{\sin kx_2 \sinh k(2h - x_2) + \sin k(2h - x_2) \sinh kx_2\} \sin \omega t \\ & +[\{\cos kx_2 \cosh k(2h - x_2) + \cos k(2h - x_2) \cosh kx_2\} \cos \omega t]. \end{aligned}$$

INDEX

Engineering

DE RE METALLICA, Georgius Agricola. The famous Hoover translation of greatest treatise on technological chemistry, engineering, geology, mining of early modern times (1556). All 289 original woodcuts. 638pp. 6¾ x 11. 0-486-60006-8

FUNDAMENTALS OF ASTRODYNAMICS, Roger Bate et al. Modern approach developed by U.S. Air Force Academy. Designed as a first course. Problems, exercises. Numerous illustrations. 455pp. 5⅜ x 8½. 0-486-60061-0

DYNAMICS OF FLUIDS IN POROUS MEDIA, Jacob Bear. For advanced students of ground water hydrology, soil mechanics and physics, drainage and irrigation engineering and more. 335 illustrations. Exercises, with answers. 784pp. 6⅛ x 9¼. 0-486-65675-6

THEORY OF VISCOELASTICITY (Second Edition), Richard M. Christensen. Complete consistent description of the linear theory of the viscoelastic behavior of materials. Problem-solving techniques discussed. 1982 edition. 29 figures. xiv+364pp. 6⅛ x 9¼. 0-486-42880-X

MECHANICS, J. P. Den Hartog. A classic introductory text or refresher. Hundreds of applications and design problems illuminate fundamentals of trusses, loaded beams and cables, etc. 334 answered problems. 462pp. 5⅜ x 8½. 0-486-60754-2

MECHANICAL VIBRATIONS, J. P. Den Hartog. Classic textbook offers lucid explanations and illustrative models, applying theories of vibrations to a variety of practical industrial engineering problems. Numerous figures. 233 problems, solutions. Appendix. Index. Preface. 436pp. 5⅜ x 8½. 0-486-64785-4

STRENGTH OF MATERIALS, J. P. Den Hartog. Full, clear treatment of basic material (tension, torsion, bending, etc.) plus advanced material on engineering methods, applications. 350 answered problems. 323pp. 5⅜ x 8½. 0-486-60755-0

A HISTORY OF MECHANICS, René Dugas. Monumental study of mechanical principles from antiquity to quantum mechanics. Contributions of ancient Greeks, Galileo, Leonardo, Kepler, Lagrange, many others. 671pp. 5⅜ x 8½. 0-486-65632-2

STABILITY THEORY AND ITS APPLICATIONS TO STRUCTURAL MECHANICS, Clive L. Dym. Self-contained text focuses on Koiter postbuckling analyses, with mathematical notions of stability of motion. Basing minimum energy principles for static stability upon dynamic concepts of stability of motion, it develops asymptotic buckling and postbuckling analyses from potential energy considerations, with applications to columns, plates, and arches. 1974 ed. 208pp. 5⅜ x 8½. 0-486-42541-X

METAL FATIGUE, N. E. Frost, K. J. Marsh, and L. P. Pook. Definitive, clearly written, and well-illustrated volume addresses all aspects of the subject, from the historical development of understanding metal fatigue to vital concepts of the cyclic stress that causes a crack to grow. Includes 7 appendixes. 544pp. 5⅜ x 8½. 0-486-40927-9

ROCKETS, Robert Goddard. Two of the most significant publications in the history of rocketry and jet propulsion: "A Method of Reaching Extreme Altitudes" (1919) and "Liquid Propellant Rocket Development" (1936). 128pp. 5⅜ x 8½. 0-486-42537-1

STATISTICAL MECHANICS: PRINCIPLES AND APPLICATIONS, Terrell L. Hill. Standard text covers fundamentals of statistical mechanics, applications to fluctuation theory, imperfect gases, distribution functions, more. 448pp. 5⅜ x 8½. 0-486-65390-0

ENGINEERING AND TECHNOLOGY 1650–1750: ILLUSTRATIONS AND TEXTS FROM ORIGINAL SOURCES, Martin Jensen. Highly readable text with more than 200 contemporary drawings and detailed engravings of engineering projects dealing with surveying, leveling, materials, hand tools, lifting equipment, transport and erection, piling, bailing, water supply, hydraulic engineering, and more. Among the specific projects outlined-transporting a 50-ton stone to the Louvre, erecting an obelisk, building timber locks, and dredging canals. 207pp. 8⅜ x 11¼. 0-486-42232-1

THE VARIATIONAL PRINCIPLES OF MECHANICS, Cornelius Lanczos. Graduate level coverage of calculus of variations, equations of motion, relativistic mechanics, more. First inexpensive paperbound edition of classic treatise. Index. Bibliography. 418pp. 5⅜ x 8½. 0-486-65067-7

PROTECTION OF ELECTRONIC CIRCUITS FROM OVERVOLTAGES, Ronald B. Standler. Five-part treatment presents practical rules and strategies for circuits designed to protect electronic systems from damage by transient overvoltages. 1989 ed. xxiv+434pp. 6⅛ x 9¼. 0-486-42552-5

ROTARY WING AERODYNAMICS, W. Z. Stepniewski. Clear, concise text covers aerodynamic phenomena of the rotor and offers guidelines for helicopter performance evaluation. Originally prepared for NASA. 537 figures. 640pp. 6⅛ x 9¼. 0-486-64647-5

INTRODUCTION TO SPACE DYNAMICS, William Tyrrell Thomson. Comprehensive, classic introduction to space-flight engineering for advanced undergraduate and graduate students. Includes vector algebra, kinematics, transformation of coordinates. Bibliography. Index. 352pp. 5⅜ x 8½. 0-486-65113-4

HISTORY OF STRENGTH OF MATERIALS, Stephen P. Timoshenko. Excellent historical survey of the strength of materials with many references to the theories of elasticity and structure. 245 figures. 452pp. 5⅜ x 8½. 0-486-61187-6

ANALYTICAL FRACTURE MECHANICS, David J. Unger. Self-contained text supplements standard fracture mechanics texts by focusing on analytical methods for determining crack-tip stress and strain fields. 336pp. 6⅛ x 9¼. 0-486-41737-9

STATISTICAL MECHANICS OF ELASTICITY, J. H. Weiner. Advanced, self-contained treatment illustrates general principles and elastic behavior of solids. Part 1, based on classical mechanics, studies thermoelastic behavior of crystalline and polymeric solids. Part 2, based on quantum mechanics, focuses on interatomic force laws, behavior of solids, and thermally activated processes. For students of physics and chemistry and for polymer physicists. 1983 ed. 96 figures. 496pp. 5⅜ x 8½. 0-486-42260-7

CATALOG OF DOVER BOOKS

A TREATISE ON ELECTRICITY AND MAGNETISM, James Clerk Maxwell. Important foundation work of modern physics. Brings to final form Maxwell's theory of electromagnetism and rigorously derives his general equations of field theory. 1,084pp. 5⅜ x 8½. Two-vol. set. Vol. I: 0-486-60636-8 Vol. II: 0-486-60637-6

QUANTUM MECHANICS: PRINCIPLES AND FORMALISM, Roy McWeeny. Graduate student-oriented volume develops subject as fundamental discipline, opening with review of origins of Schrödinger's equations and vector spaces. Focusing on main principles of quantum mechanics and their immediate consequences, it concludes with final generalizations covering alternative "languages" or representations. 1972 ed. 15 figures. xi+155pp. 5⅜ x 8½. 0-486-42829-X

INTRODUCTION TO QUANTUM MECHANICS With Applications to Chemistry, Linus Pauling & E. Bright Wilson, Jr. Classic undergraduate text by Nobel Prize winner applies quantum mechanics to chemical and physical problems. Numerous tables and figures enhance the text. Chapter bibliographies. Appendices. Index. 468pp. 5⅜ x 8½. 0-486-64871-0

METHODS OF THERMODYNAMICS, Howard Reiss. Outstanding text focuses on physical technique of thermodynamics, typical problem areas of understanding, and significance and use of thermodynamic potential. 1965 edition. 238pp. 5⅜ x 8½. 0-486-69445-3

THE ELECTROMAGNETIC FIELD, Albert Shadowitz. Comprehensive undergraduate text covers basics of electric and magnetic fields, builds up to electromagnetic theory. Also related topics, including relativity. Over 900 problems. 768pp. 5⅜ x 8¼. 0-486-65660-8

GREAT EXPERIMENTS IN PHYSICS: FIRSTHAND ACCOUNTS FROM GALILEO TO EINSTEIN, Morris H. Shamos (ed.). 25 crucial discoveries: Newton's laws of motion, Chadwick's study of the neutron, Hertz on electromagnetic waves, more. Original accounts clearly annotated. 370pp. 5⅜ x 8½. 0-486-25346-5

EINSTEIN'S LEGACY, Julian Schwinger. A Nobel Laureate relates fascinating story of Einstein and development of relativity theory in well-illustrated, nontechnical volume. Subjects include meaning of time, paradoxes of space travel, gravity and its effect on light, non-Euclidean geometry and curving of space-time, impact of radio astronomy and space-age discoveries, and more. 189 b/w illustrations. xiv+250pp. 8⅜ x 9¼. 0-486-41974-6

STATISTICAL PHYSICS, Gregory H. Wannier. Classic text combines thermodynamics, statistical mechanics and kinetic theory in one unified presentation of thermal physics. Problems with solutions. Bibliography. 532pp. 5⅜ x 8½. 0-486-65401-X

Paperbound unless otherwise indicated. Available at your book dealer, online at **www.doverpublications.com**, or by writing to Dept. GI, Dover Publications, Inc., 31 East 2nd Street, Mineola, NY 11501. For current price information or for free catalogues (please indicate field of interest), write to Dover Publications or log on to **www.doverpublications.com** and see every Dover book in print. Dover publishes more than 500 books each year on science, elementary and advanced mathematics, biology, music, art, literary history, social sciences, and other areas.